Drawing upon Nature: Studies for the Blaschkas' Glass Models

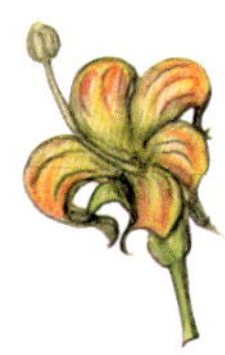

Selected and Introduced
by Susan M. Rossi-Wilcox
and David Whitehouse

Cover:
Bromelia pinguin *L.*
Rat Pineapple
Bromeliaceae
Model 451 (1894)

Editor: Richard W. Price
Design and Typography: Jacolyn S. Saunders
Reference Librarian: Gail P. Bardhan
Rights and Reproductions: Jill Thomas-Clark
Proofreader: Mary Chervenak

Standard Book Number: 0-87290-166-1
Library of Congress Control Number: 2007920268

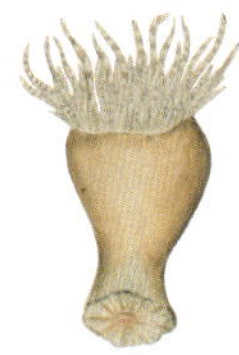

Contents

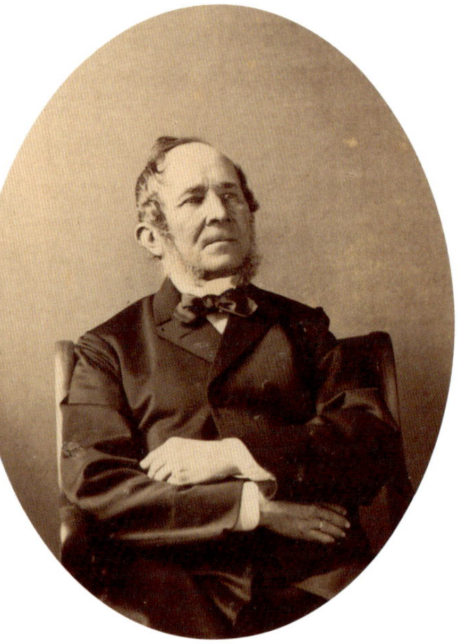

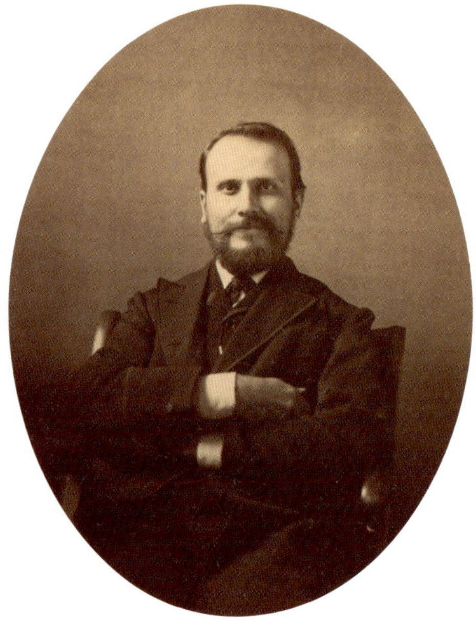

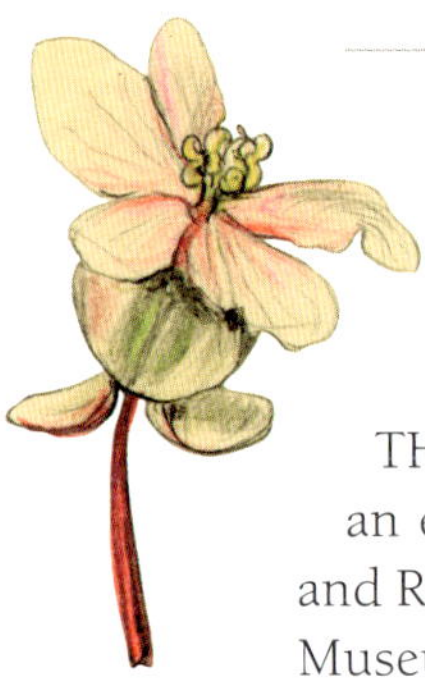

Preface

THIS BOOK was designed to supplement an exhibition of glass models by Leopold and Rudolf Blaschka on view at The Corning Museum of Glass from May 18 to November 25, 2007. It was conceived, however, when the Corning Museum and the Botanical Museum of Harvard University jointly purchased the remaining Blaschka studio materials from Gertrud Pones, Mrs. Rudolf (Frieda) Blaschka's niece, in 1993. From the time we saw the reference materials on which the Blaschkas relied in constructing their models, we have wanted to provide at least a glimpse of the drawings in a publication. The book contains reproductions of 59 of the more than 900 drawings of invertebrate animals and plants created by the Blaschkas,[1] which are now among the treasures of the Corning Museum's Rakow Research Library.

Despite the meticulous and "finished" appearance of these drawings, the Blaschkas never prided themselves as draftsmen, and the drawings were simply a means to an end. Instead, the two men's claim to fame derives from the amazing glass models of soft-bodied animals and delicate plants that they created. The Blaschkas' models represent more than 1,500 species of invertebrate animals and plants. They are surrogates for the real things, and they have served as teaching aids for generations of students and teachers of natural history. With few exceptions, these models are of extraordinary accuracy and beauty.

Leopold Blaschka (Fig. 1; left) was born on May 27, 1822, in Böhmisch Aicha, in what is now the Czech Republic, and he died in Dresden on July 3, 1895. He married Karolina Riegel on November 6, 1854. Their son, Rudolf (Fig. 2; right), was born on June 17, 1857. Rudolf married Martha Frieda Richter on May 16, 1911, but they did not have any children to continue the family's glassworking traditions. Rudolf died on May 1, 1939.

1. It also presents six plates from printed books and journals that were used by the Blaschkas as sources of information about invertebrate animals.

Figure 1
Leopold Blaschka, 1822–1895. Rakow Research Library, The Corning Museum of Glass.

Figure 2
Rudolf Blaschka, 1857–1939. Rakow Research Library, The Corning Museum of Glass.

Leopold began to make models of invertebrates in 1863, and, assisted from 1876 by his son, he continued to produce them until 1890. (Leopold also made a few botanical models for his own pleasure in the early 1860s, but he concentrated commercially on invertebrate animals.) In this 27-year period, the Blaschkas created thousands of models of soft-bodied animals, which they sold to museums and universities in Europe, North America, the Far East, and Australasia.

With the assistance of his son, Leopold Blaschka returned to the making of botanical models in 1886. He constructed these models for the Botanical Museum of Harvard University. Four years later, he and Rudolf signed an exclusive contract with Harvard to make plant models full-time. After Leopold's death, Rudolf continued to produce botanical models for the remainder of his career. Together, this father-and-son team turned out roughly 4,300 plant models, including nearly 850 life-size reproductions.

The Blaschkas fashioned their models by flameworking. They heated glass tubes and rods in a flame until they became soft and could be pulled into different shapes with tweezers, cut into the required sizes with scissors, and attached to one another by fusion. Initially, the Blaschkas softened the glass in the flame of a paraffin lamp, then in an alcohol burner. Later, Rudolf used a gas-burning torch, the forerunner of the propane- and oxygen-burning torches employed by flameworkers today.

Although the essential ingredient of every model was colorless or colored glass, the Blaschkas used a variety of other materials: copper wire to make armatures, animal glue to assemble complex models, watercolor and oil-based paint, and even paper to represent the internal organs of certain invertebrates.

Although the Blaschkas used materials other than glass, flameworkers today revere them as absolute masters of their craft. They admire and sometimes shake their head in disbelief at the ability of Leopold and Rudolf Blaschka to observe the smallest, most ephemeral details of the natural world and to capture them permanently in glass.

From the time the models of glass flowers were first displayed publicly, they have been regarded as scientifically accurate. These claims can be validated by examining the Blaschkas' drawings. Likewise, as zoology professors may attest, the best of the detailed invertebrate models are more realistic than preserved specimens, and they rival three-dimensional replicas made of any other material, including plastic and fiberglass, that are commonly sold for classroom instruction today.

From a curatorial perspective, the Blaschkas' models provide a running commentary on the development of the biological sciences in the late 19th and early 20th centuries, ranging from a growing interest in embryology to concerns about the blight of fruit crops. In addition to staying abreast of scientific advances, Leopold and Rudolf were astute businessmen who could readily identify commercially viable opportunities. These qualities, combined with their consummate craftsmanship, resulted in models that provide a vivid legacy found in museum collections worldwide.

In October 2006, some 60 historians of science, biologists, and curators (including the authors of this book) met in Dublin, Ireland, to participate in the first international congress devoted to the Blaschkas and their work. Among the key speakers was Henri Reiling of Utrecht, the Netherlands, who shared his unrivaled knowledge of the history of the invertebrate models. We thank him for more than a decade of cooperation in studying the Blaschkas. (Indeed, much of the information contained in the essay on invertebrates first appeared in his publications listed on page 160.) The Dublin program included lectures, panels, and a visit to the collection of 530 models of invertebrates acquired by the National Museum of Ireland–Natural History between 1878 and 1888. The second international Blaschka congress will expand on the research reported in Dublin when it convenes in Corning, New York, in October 2007.

The two essays that follow are styled differently. While David Whitehouse's treatment of the models of invertebrate animals is a relatively straightforward historical account, the section on botanical models, written by Susan M. Rossi-Wilcox, provides a travelogue of Rudolf Blaschka's field expeditions to North America in 1892 and 1895. Her essay reconstructs his ambitious itinerary, notes his impressions as a field botanist, and reports on the problems he encountered along the way. It was during these trips that Rudolf created the meticulously rendered drawings that would serve as the basis for the glass models he would later produce in his Dresden studio.

Following the essays are two collections of drawings, 27 (and three prints) of invertebrates and 30 of botanical specimens, with captions that identify and describe the species represented. The brief introduction to the drawings offers the reader a guide for interpreting the Blaschkas' references.

Susan M. Rossi-Wilcox
Curatorial Associate
Botanical Museum–Harvard University Herbaria

David Whitehouse
Executive Director
The Corning Museum of Glass

Models of Invertebrate Animals (1863–1890)

David Whitehouse, Executive Director, The Corning Museum of Glass

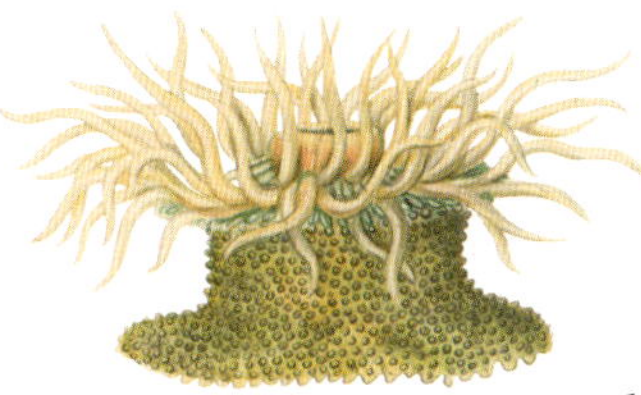

THE STORY of the Blaschkas begins in the small town of Böhmisch Aicha (now Český Dub in the Czech Republic), where Leopold's father continued the family tradition of flameworking. When Leopold was a student, his favorite subjects were natural history and painting, and a visiting artist urged him to pursue a career in art. However, after serving an apprenticeship as a goldsmith and gem cutter in nearby Turnau (now Turnov), Leopold entered the family business of making costume jewelry and other fancy goods out of metal and flameworked glass. The catalog of its displays at London's Great Exhibition in 1851 gives an indication of the range of the wares:

> BLASCHKA & SONS, *Liebenau, Bohemia*—Manufacturers.
> Paste, for artificial precious stones, beads, glass buttons, lustre pendants, articles in pinchbeck, &c.[1]

Leopold suffered two devastating losses. In 1850, his first wife died of cholera. His father died two years later. In 1853, the grieving Leopold took time off and visited the United States. On the outward journey, his ship was becalmed for two weeks near the Azores, and Leopold passed the time collecting and drawing jellyfish and other marine invertebrates. Apart from illustrations in books, he had never before seen such animals, and their glasslike transparency fascinated him.

Leopold's description of what he saw (in a translation kindly provided by Henri Reiling) conveys his sense of wonder—a sense that he later expressed in his glass models. Here, he describes phosphorescent creatures observed after dark:

> It is a beautiful night in May. Hopeful, we look out over the darkness of the sea, which is as smooth as a mirror; there emerges all around in various places a flashlike bundle of light beams, as if it is surrounded by thousands of sparks, that form true bundles of fire and of other bright lighting spots, and the seemingly mirrored stars. There emerges close before us a small spot in a sharp greenish light, which becomes ever larger and larger and finally becomes a bright shining sunlike figure.

Despite his love of marine invertebrates, the first glass models that Leopold made, following his return to Europe, were of plants, not animals. After the death of his wife, he had sought consolation in collecting, studying, and painting plants, and in the late 1850s, he began to create models of plants for his own amusement. These models came to the attention of Prince Camille de Rohan, an aristocratic horticulturist who had established a world-famous garden on one of his estates, at Sychrov Castle, not far from Aicha. Between 1860 and 1862, the prince exhibited about 100 models of orchids and other exotic plants, which he displayed on two artificial tree trunks in his palace in Prague. The orchids do not survive, but we suspect that the bouquet illustrated in Figure 1 is a rare example of Leopold's early experiments with models of plants. (The bouquet is completely unlike the botanical models made for Harvard in and after 1887, and the colored glass from which it was made is closely

1. Great Exhibition (1851: London, England), *Official Descriptive and Illustrated Catalogue*, London: Spicer Brothers, 1851, v. 3, p. 1,037.

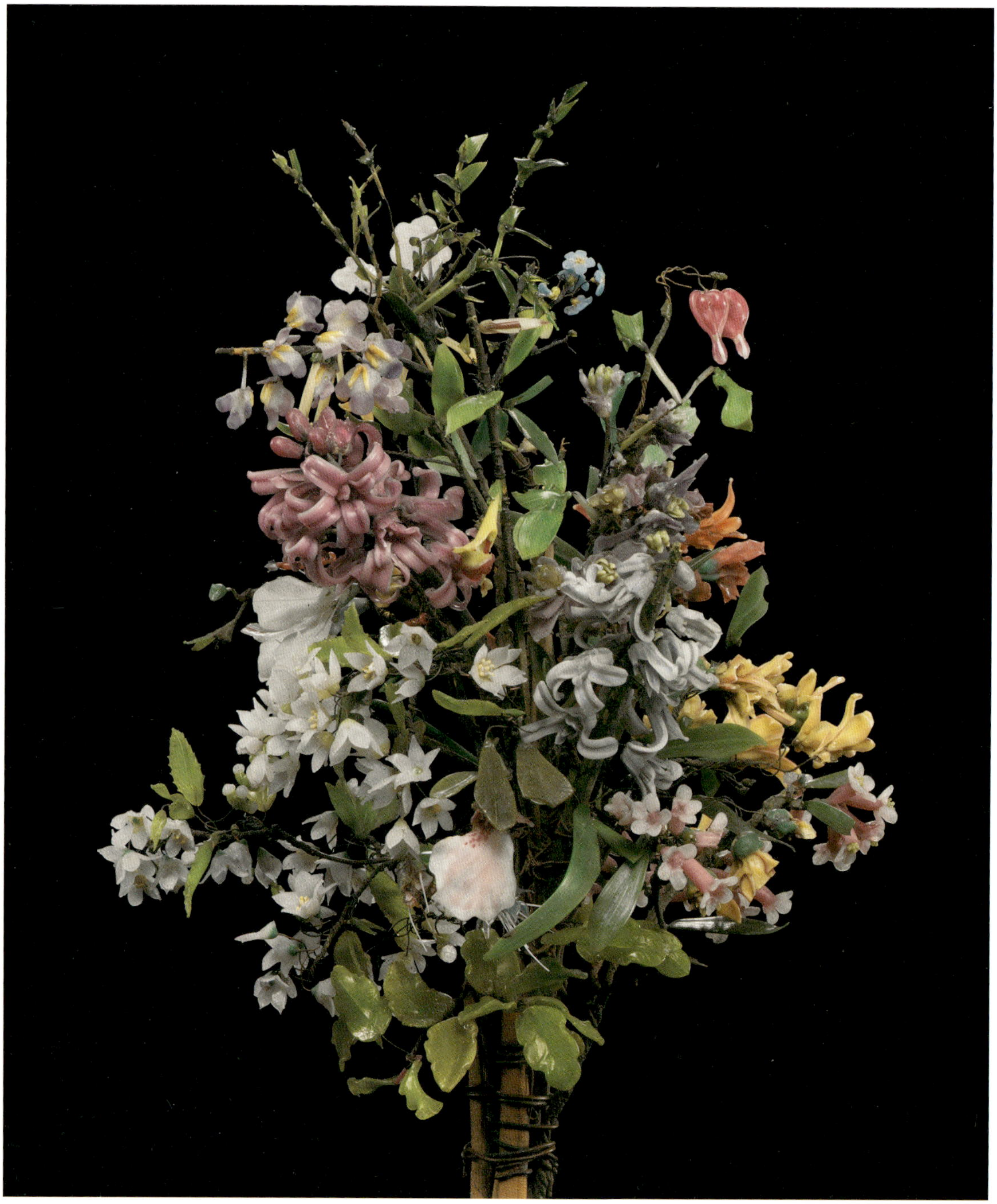

Figure 1. *Bouquet. Leopold Blaschka, probably about 1860–1865. OH. 36 cm. The Corning Museum of Glass (79.3.90).*

similar to the glass used for the earliest invertebrate models in and after 1863.)

In a letter of 1896 that is preserved at Harvard University, Rudolf Blaschka provided an important insight into his father's work, supplementing the information contained in the catalog of the Great Exhibition (see above). When Leopold entered the business, it consisted of making costume jewelry assembled from glass and metal components, metal attachments for the lusters in chandeliers, and other fancy goods. At an unknown date, the business expanded to include flameworked laboratory equipment. In or just after 1864, Leopold was commissioned to produce brooches and other jewelry decorated with flameworked flowers: forget-me-nots, violets, and roses. He also produced flameworked glass eyes in various colors and sizes. Some of these were for people who had lost an eye, and others were for taxidermists who supplied the growing demands of natural history museums for animals whose skins were preserved and mounted for display. Leopold's production of jewelry ceased about 1873, but the manufacture of glass eyes and scientific apparatus continued until the 1880s.

The prince also gave Leopold an introduction to Prof. Ludwig Reichenbach, director of both the botanical garden and the natural history museum in Dresden. Subsequently, Leopold's models were exhibited in that city's botanical garden, and in a museum in Liège, Belgium, where they were destroyed in a fire.

Leopold now moved his family from Aicha to Dresden, and, in 1863, Reichenbach commissioned him to make models of sea anemones to be displayed in the natural history museum. These glass anemones attracted the attention of other museum directors, and Reichenbach predicted that Leopold had a promising future as a maker of scientific models.

Reichenbach's prediction proved to be correct. Indeed, such was Leopold's success that, at the age of 41, he turned his back on the other business and devoted most of his energy to the making of models of invertebrate animals.

It is easy to see why Leopold became a successful modelmaker. He was, after all, a gifted flameworker. But this is only part of the explanation. Other factors were also at work. Leopold's apprenticeship as a jeweler and his years in the family business had provided him with opportunities to work on a very small scale, and these experiences may also have given him the idea that complex models (such as bouquets and, later, invertebrate animals) could be assembled like jewelry.

Leopold used several design sources. The first of these was the printed page. He made his earliest models of invertebrates by copying the illustrations in the most up-to-date textbooks of zoology. One was *Actinologia Britannica: A History of the British Sea-Anemones and Corals*, published between 1858 and 1860 by the self-taught British naturalist Philip Henry Gosse (1810–1888). Gosse's illustrations provided Leopold with images of sea anemones, but they also suggested that the models could be displayed on rocklike bases, which Leopold copied immediately (Figs. 2–5).

Gosse's plates proved to be a mixed blessing, however. The anemones were not shown at a uniform scale; indeed, the plates themselves gave no indication of scale. Moreover, the plates were open to misinterpretation, and if Gosse illustrated a specimen as seen from above, Leopold produced a flat, disklike model.

In an earlier book, *The Aquarium* (1854), Gosse had popularized the notion of keeping sea creatures in oxygenated saltwater aquariums (then a novelty), and this soon allowed Leopold to nurture and copy living specimens as well as illustrations in books.

Gosse had a powerful voice, and not only as a naturalist. In 1857, he published the strangely titled *Omphalos/Creation*, in which he expounded the view that, at the Creation, God had created all things for all time. Gosse was thus at odds with Charles Darwin (1809–1882), whose theory of evolution would be published two years later in *On the Origin of Species*.

A more distinguished naturalist who inspired Leopold was Ernst Haeckel (1834–1919), an eminent but controversial zoologist (and champion of Darwinism) at the University of Jena. Haeckel was a prodigious researcher who made major advances in our understanding of marine biology. He studied and published specimens collected by the Challenger Expedition, an important feat of oceanographic exploration carried out by the British Admiralty and the Royal Society between 1872 and 1876. During this period, H.M.S. *Challenger*, a 2,300-ton corvette, crisscrossed the world's oceans, covering nearly 69,000 nautical miles. While the expedition's oceanographers mapped the seabed and documented currents and temperatures, biologists collected thousands of species of marine life. The find-

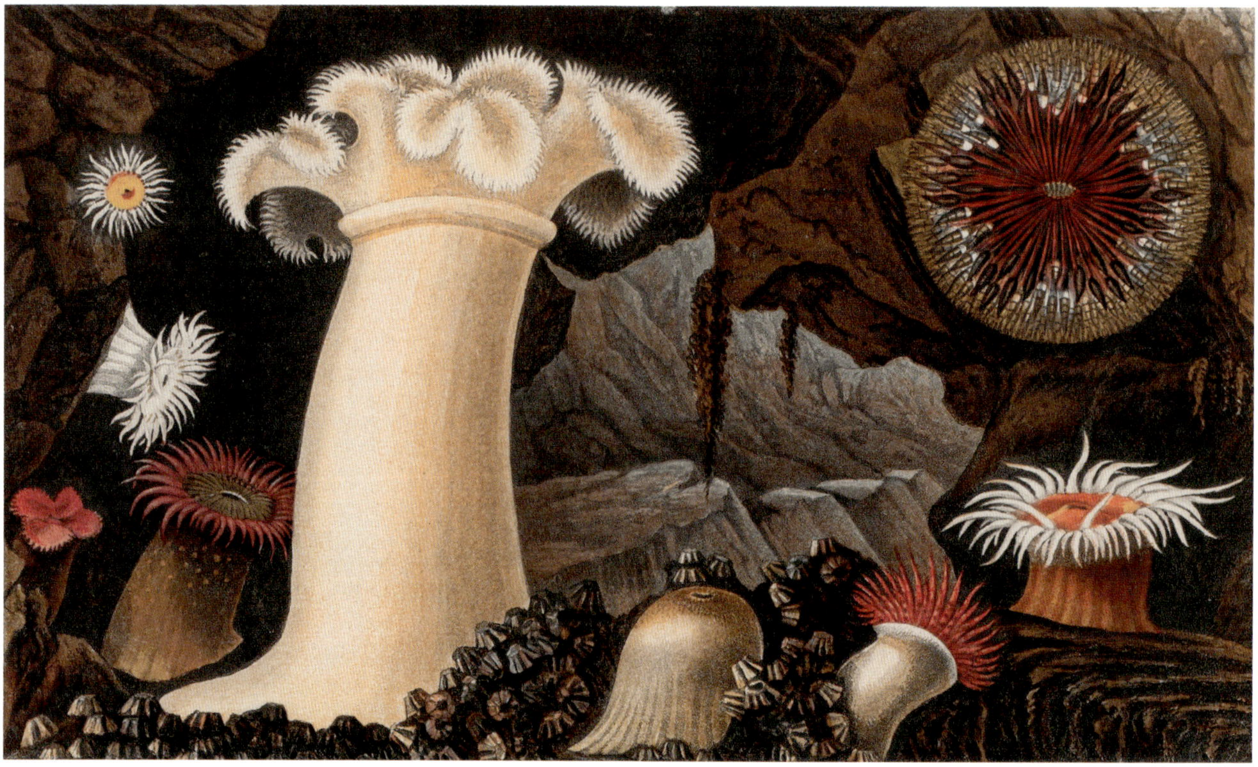

Figure 2. *Sea anemones, including (no. 1, largest)* Actinoloba dianthus, *shown in plate I of Philip Henry Gosse,* Actinologia Britannica: A History of the British Sea-Anemones and Corals, *London: Van Voorst, 1860.*

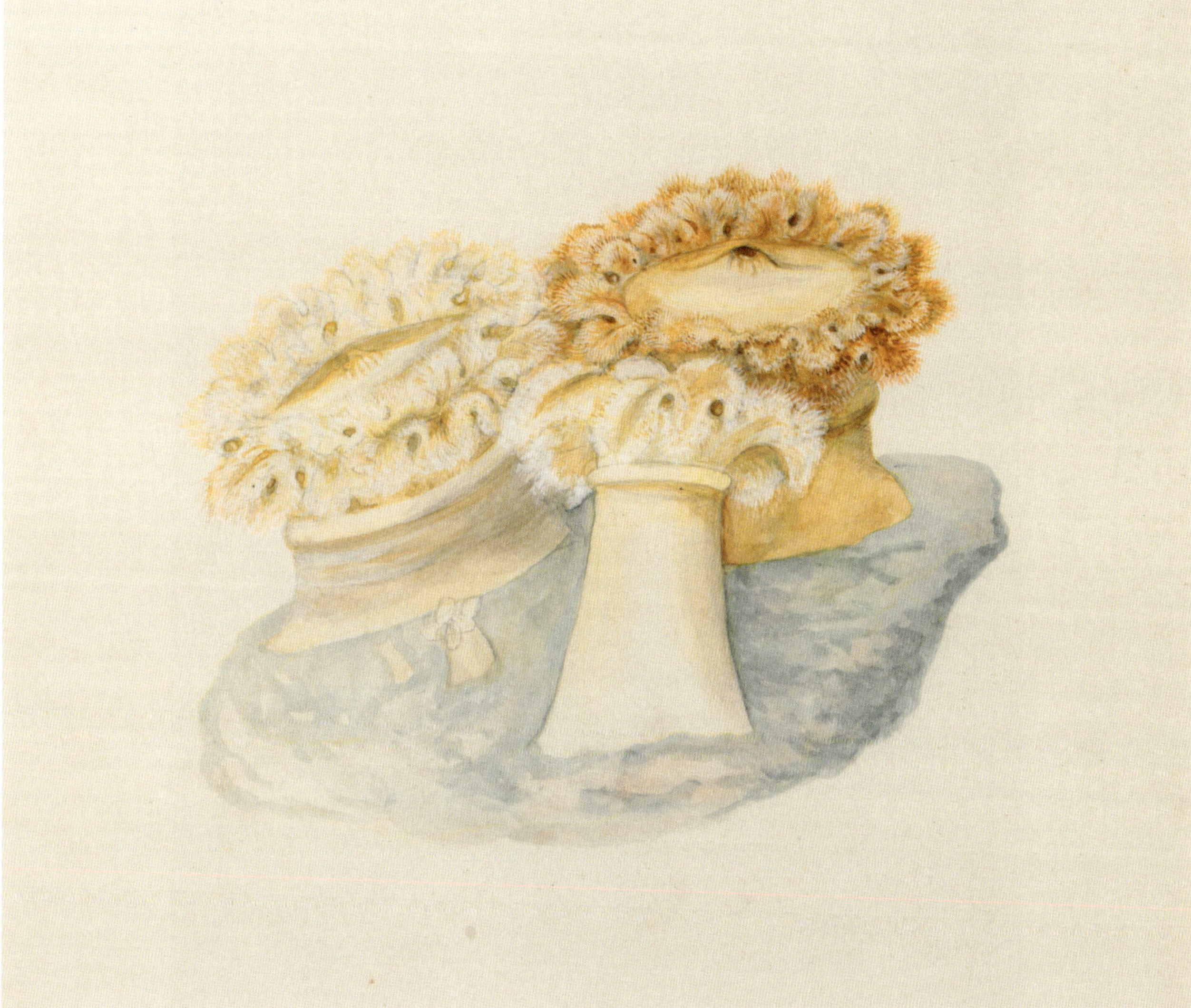

Figure 3. *Sea anemones, including* Actinoloba dianthus, *copied by Leopold Blaschka from Gosse's* Actinologia Britannica. *Pencil, ink, and watercolor on paper, 39 cm x 33 cm. Rakow Research Library, The Corning Museum of Glass.*

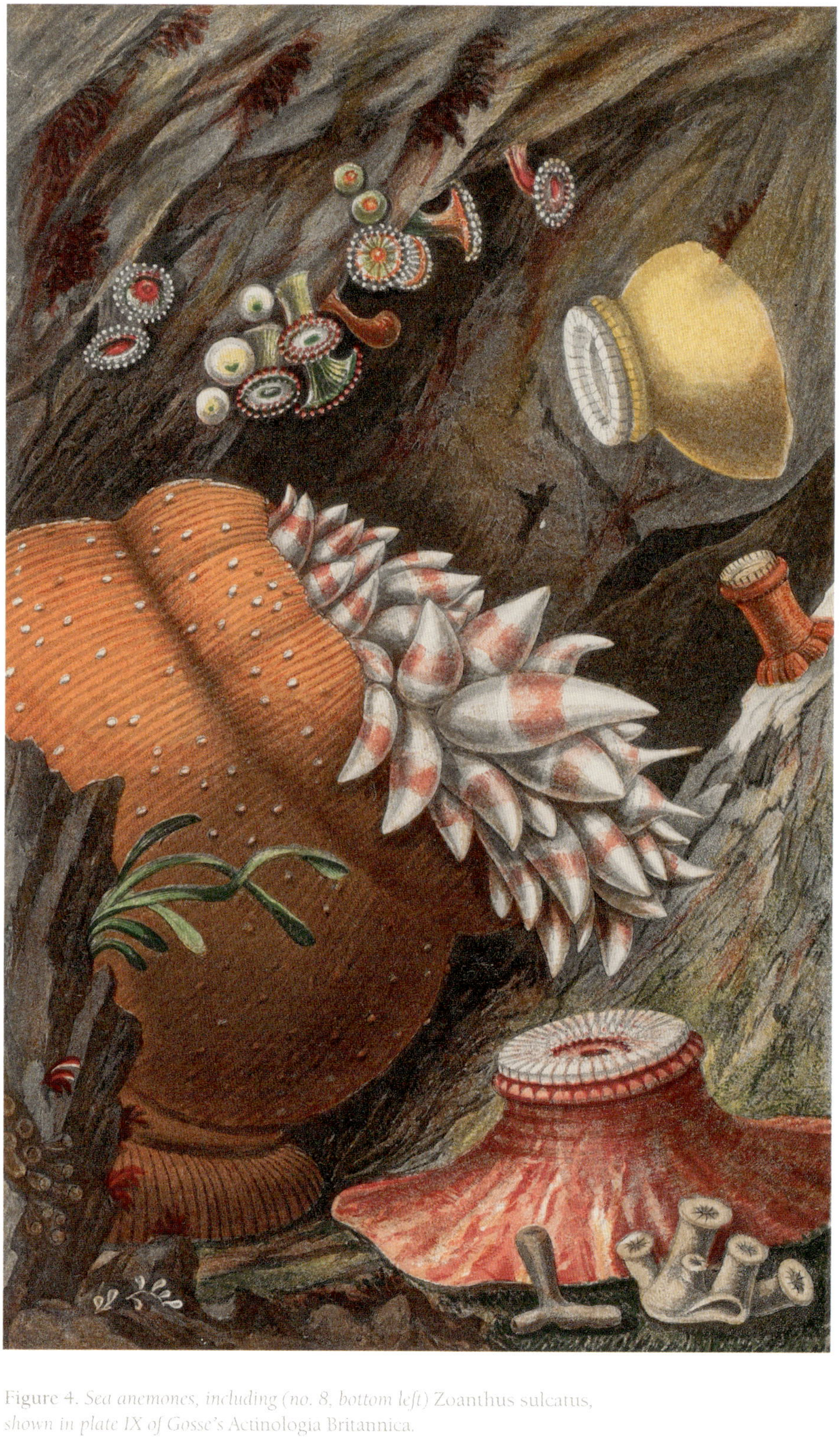

Figure 4. *Sea anemones, including (no. 8, bottom left)* Zoanthus sulcatus, *shown in plate IX of Gosse's* Actinologia Britannica.

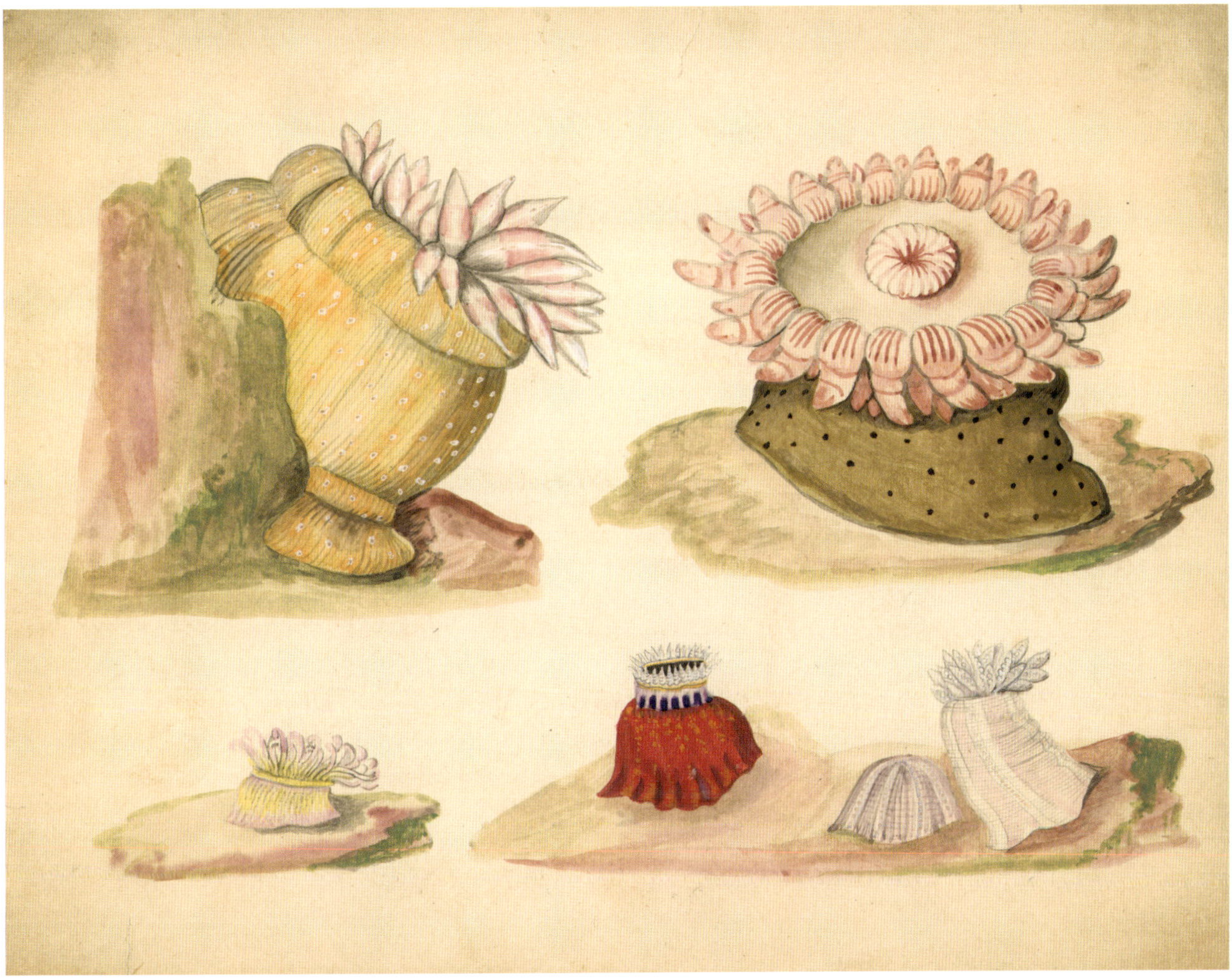

Figure 5. *Sea anemones, including* Zoanthus sulcatus, *copied by Leopold Blaschka from Gosse's* Actinologia Britannica. *Pencil, ink, and watercolor on paper, 23 cm x 29 cm. Rakow Research Library, The Corning Museum of Glass.*

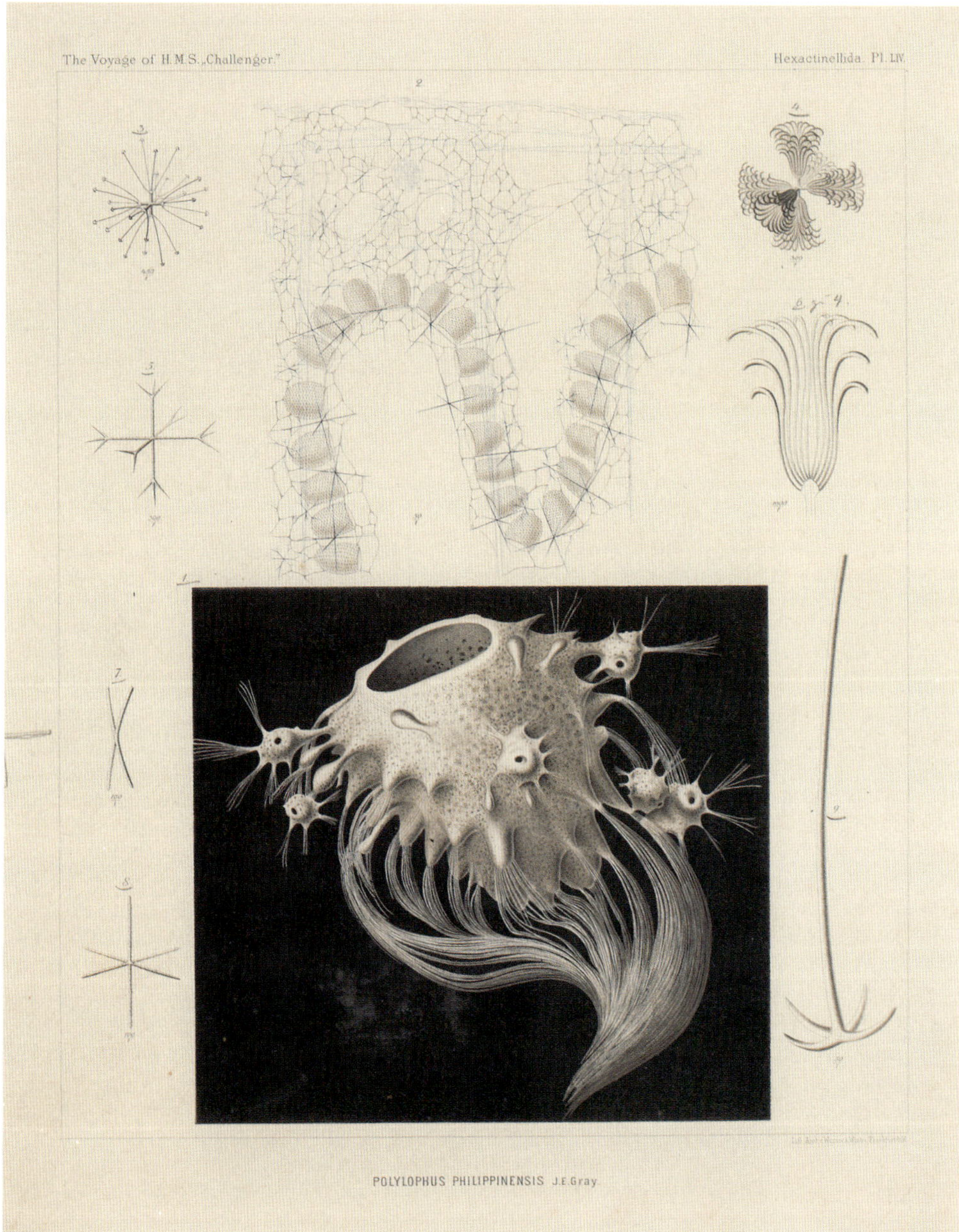

Figure 6. Polylophus philippinensis *Gray (glass sponge). This lithograph (32 cm x 26 cm) was owned by the Blaschkas. It is plate LIV in F. E. Schultz,* Report of the Scientific Research of the Voyage of H.M.S. Challenger during the Years 1873–76, *v. 21,* Hexactinellida, *London: Her Majesty's Government, 1887.*

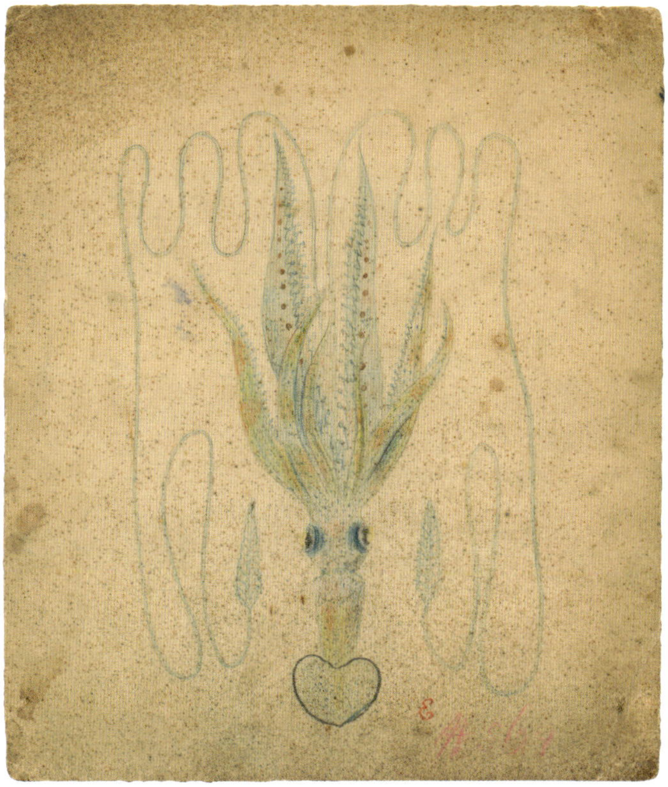

Figure 7
Loligopsis varanii *Férussac (squid). Drawing by Leopold Blaschka. Pencil, ink, and watercolor on paper, 29 cm x 24 cm. Rakow Research Library, The Corning Museum of Glass.*

Figure 8
One of the Blaschkas' models of Loligopsis varanii, *about 1885. H. 24 cm. Blaschka Marine Invertebrate Collection, Department of Ecology and Evolutionary Biology, Cornell University, Ithaca, New York.*

ings of the expedition were published in 50 volumes. Figure 6 shows a glass sponge (so called because it has siliceous spicules) published in one of the expedition's reports.

In 1877, Leopold contacted Haeckel, who lent the Blaschkas books so that they could copy the illustrations. They also purchased a copy of the first volume of Haeckel's monograph on jellyfish.

Haeckel was deeply interested in the decorative arts, and between 1899 and 1904 he published *Kunstformen der Natur* (Art forms of nature), which had a profound influence on the development of the Art Nouveau style. The inspiration of this style in natural forms is illustrated (fortuitously, since there is no direct connection) by the Blaschkas' working drawing of a squid (Fig. 7), their glass model of the same animal (Fig. 8), and such creations as the centerpiece designed by Harry Powell for Count Minerbi and made at the Whitefriars Glass Works in London in 1906 (Fig. 9).

Over the years, the Blaschkas worked from books and other publications from all over Europe. A small selection of plates from these printed materials appears on pages 10, 12, 14, 47, 67, and 73.

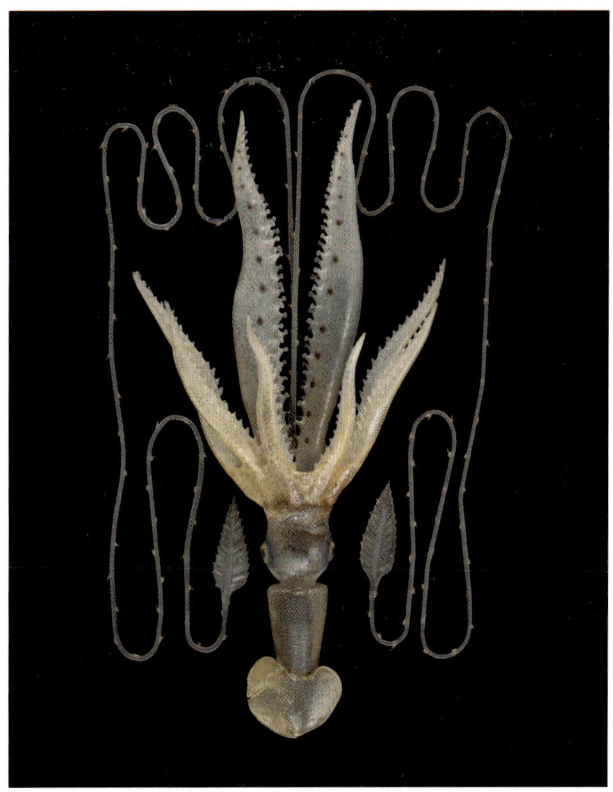

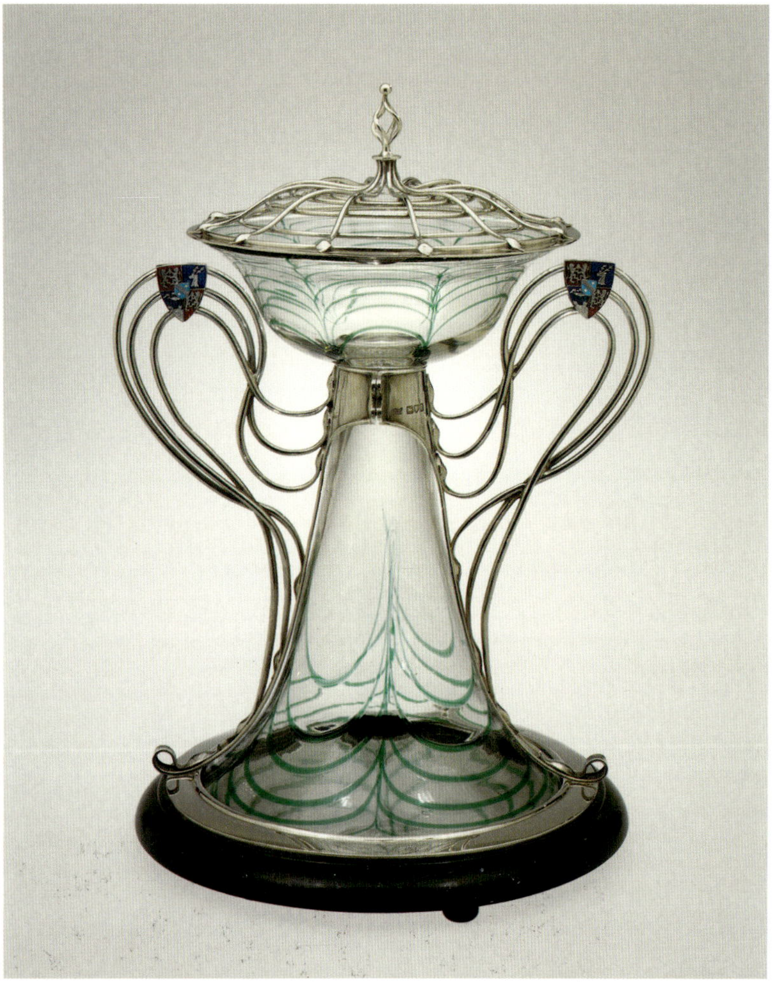

Figure 9
Centerpiece, blown, trailed and pulled decoration; wood and silver mounts, with enameled shields. England, London, James Powell and Sons, Whitefriars Glass Works, designed by Harry Powell, 1906. H. 27 cm. The Corning Museum of Glass (90.2.3, gift of Alastair Duncan).

Leopold's second design source consisted of animals preserved in glass jars of alcohol. These, too, were a mixed blessing. In time, the specimens lost their natural coloring and, having no backbones to support them, their bodies gradually subsided into more or less shapeless masses in the bottom of the jars. Nevertheless, in 1877, Leopold ordered specimens preserved in alcohol from the zoological station in Naples to compare with illustrations in books. It may be no coincidence that this departure occurred just one year after Rudolf joined his father. Rudolf was an enthusiastic assistant, and he may have injected new ideas into the business.

In any case, it was at about the same time or not long afterward that the Blaschkas began to maintain living specimens in seawater aquariums of the kind promoted by Gosse. Presumably they did not have an aquarium when they ordered preserved specimens from Naples in 1877, although letters written in 1880 note that they had had aquariums for "several" years, and that they had kept anemones alive for "one and a half" and "several" years. The Blaschkas acquired live specimens from Naples, from Chioggia and Trieste in the upper Adriatic, from Weymouth on the English Channel, and from suppliers on the coasts of the North and Baltic Seas. Their generally successful method of transporting live specimens of sea anemones and mollusks over long distances consisted of wrapping them in wet seaweed, placing them in glass jars, and packing the jars in baskets.

The Blaschkas' models of invertebrates were well timed, coinciding with the aquarium craze that had begun in England in the mid-19th century. This craze was fueled by the availability of inexpensive plate glass and the discovery that seaweed could be used to oxygenate the water in aquariums. The first large public aquarium was established by the Zoological Society of London in 1853. The largest aquarium was probably the zoological garden in Paris, which opened eight years later.

With the Blaschkas' models, it was possible to stock waterless aquariums with sea anemones and other invertebrates. These required little or no maintenance and, unlike true aquariums, no restocking after the living specimens had died. In the preface to the earliest known sales catalog of Blaschka models (about 1871, in German), Ludwig Reichenbach drew attention to the advantage of glass models over specimens preserved in alcohol: they retained both their shape and their color. The first models of anemones had soon been joined by corals, jellyfish, mollusks, and other species. As the business prospered, additional German-language catalogs were issued in 1874 and 1875 (no copy bearing the latter date is known to have survived).

The Blaschkas had one other source of information to help them make accurate models: their archive of drawings, some of which, as noted above, were copied from textbooks. Other drawings, however, were made from their own observations. In 1879, for example, Rudolf went on a field trip to the upper Adriatic, which afforded him an opportunity to observe a greater number and variety of invertebrates than he could have hoped to see in his father's aquariums. This modest expedition to Italy foreshadowed his far more ambitious field trips to the United States and the Caribbean in 1892 and 1895 (see pages 22–32). Each of the drawings is numbered, and the same numbers were used to identify the models that Leopold offered for sale in his catalogs.

The models of sea anemones and other invertebrate animals proved to be a profitable business, in part because of the Blaschkas' skill in producing minutely detailed replicas, but also because they were working during a period of momentous cultural change. That period had commenced in 1789, when the French Revolution caused the demise of one of the foremost states in Europe and the collapse of its social and religious institutions. The aftershocks of the Revolution were felt all over Europe. Traditional values were challenged and, in many cases, transformed. At least two of these transformations affected the Blaschkas directly: the development of science and the expansion of public education.

In science, the Swedish botanist Carolus Linnaeus (1707–1778) had put the classification of animals and plants on a systematic basis during the second and third quarters of the 18th century, and his work was carried forward in the late 18th and early 19th centuries by biologists such as Georges Cuvier (1769–1832) and Jean-Baptiste-Pierre-Antoine de Monet de Lamarck (1744–1829). Their research led to a flood of publications. Some of them, such as Pierre Joseph Redouté's *Les Liliacées* (published between 1802 and 1816) and *Les Roses* (1817–1824) and John James Audubon's *Birds of America* (1827–1838), were not only scientific treatises but also masterpieces of draftsmanship and printing.

In public education, museums were springing up like mushrooms in the mid-19th century. Before 1800, almost all European and American museums were private. The Louvre in Paris did not open its doors to the public until 1793, and the Prado in Madrid followed suit in 1809. Subsequently, public museums were established all over Europe, as well as in the United States and other countries colonized by European powers. The Smithsonian Institution in Washington, D.C., founded in 1846, is a prime example. Unlike most of the earlier museums, which focused on "curiosities" or the fine and decorative arts, the Smithsonian and other 19th-century foundations included collections that illustrated the natural sciences, and many of them displayed biological specimens arranged according to the classifications of Linnaeus and his successors.

The Blaschkas succeeded because their models solved a problem that confronted all directors of natural history museums. While the vertebrates (that is, animals, birds, and fish with backbones) were relatively easy to display—live specimens were killed, and their skins were stuffed and mounted in more or less lifelike poses—few of the invertebrates (jellyfish, squids, and so forth) could be exhibited in this manner. Indeed, the only method of displaying invertebrates was in bottles of alcohol. This limited the scope of what could be exhibited, and, in any case, the animals eventually collapsed and their colors faded. The Blaschkas' models provided curators with displays featuring forms and colors that were permanent.

The proliferation of museums and the desire to display comprehensive natural history collections led to the emergence of a new profession: the supplier of fossils and zoological specimens. One such supplier was Robert Damon of Weymouth, England. He was typical of the new natural history dealers: adventurous, well informed, and served by a global network of agents. His obituary in the London *Times* of May 7, 1889, records that he "traveled all through Europe and penetrated into Asia and Africa. . . . When in Russia, he traveled . . . down the Volga . . . for the purpose of collecting the fishes of the Caspian Sea." Damon was also a serious scholar. He wrote a handbook of the geology of the Weymouth region, and his collection of fossils is now in the Natural History Museum in London.

In the early years of his business, Leopold had realized that dealers such as Damon provided excellent outlets for his models. They were in touch with the curators Leopold was trying to reach, and their networks offered him the possibility of selling his models all over Europe and beyond. The first dealer who is known to have handled Leopold's models was Václav Frič of Prague, now the capital of the Czech Republic. Frič went into business in 1862, and he supplied natural history specimens to an international clientele. In 1866, for example, he sold models to the museum in Leeds, England. Not long afterward, Leopold appointed Damon to be his agent in the United Kingdom. In a letter to a British customer, Leopold explained that it was more efficient for a local agent to collect orders and payments, and to send these to Dresden in batches, than for him to process them individually.

Leopold's strategy succeeded, and the extent of his business abroad is revealed by an English-language trade catalog, the manuscript of which survives in the Rakow Research Library at The Corning Museum of Glass. In the preface, Leopold wrote:

> [Models of invertebrate animals] have been purchased by . . . museums and scholastic establishments in all the quarters of the globe . . . in New Zealand . . . in Tokio [*sic*], Japan . . . for the Indian Museum in Calcutta . . . in the United States of America by Professor Ward's Natural Science Establishment in Rochester, New York; for the Museum of Comparative Zoology in Cambridge, Massachusetts; for the Boston Society of Natural History; the University of Cornell; the Welleslev Female College. . . . In Great Britain, Scotland and Ireland, copies have been conveyed to London, Edinburgh and Dublin. . . . In Austria, orders have not only been made for the Imperial Royal Court collection, but also for the universities in Innsbruck, Graz, Czernovitz, and so forth. In Germany, purchases have been made for the universities of Berlin, Bonn, Koenigsberg, Jena, Leipzig, Rostock and many other museums.

Leopold's claim to supply "all the quarters of the globe" was no exaggeration. In addition to the places just mentioned, his order book for the years between 1883 and 1890, preserved in the Rakow Library, records sales to agents and individual customers in France, Belgium, the Netherlands, Italy, Switzerland, and Australia.

The Blaschkas' output was prodigious. The South Kensington Museum in London (now the Natural History Museum), for example, acquired 784 models in 1876; Cornell University in Ithaca, New York, purchased some 600 models; the Museum of Natural History in Dublin acquired 530 models between 1878 and 1886; the Museum of Comparative Zoology at Harvard University bought about 350 models; and the Boston Museum of Science has 311 models.

As the buyers' names suggest, colleges in which natural history was taught were proliferating almost as quickly as public museums of natural history, and the acquisition of teaching collections became an urgent necessity. Under the charismatic leadership of the zoologist Louis Agassiz (1807–1873), Harvard University began to form teaching collections in the 1850s. Agassiz founded the Museum of Comparative Zoology in 1858, and, at its opening, he made an impassioned speech in which he complained that American students who wished to study the natural sciences in depth were compelled to visit Europe because of the lack of teaching collections in the United States. He vowed to remedy the situation by bringing to America comprehensive collections of zoological and geological specimens. Indeed, it was Agassiz who acquired the zoological museum's 350 Blaschka models.

One of Agassiz's assistants at Harvard was Henry Augustus Ward, a young American with a passion for geology. Like Damon, Ward was an adventurer. He left Harvard to take advantage of an expense-paid visit to France. This was the first of many trips abroad during which he collected fossils and other natural history specimens. In France, he was befriended by Madame Clicquot (wine lovers remember her as Veuve Clicquot), who allowed him to collect fossils in her vineyards in Champagne. In England, he sold his French fossils to the Natural History Museum in London. He then collected English fossils and sold them in France.

Once, when returning from Africa, Ward stopped in London, where he was invited to a reception in honor of David Livingstone. Unfortunately, Ward did not have a tuxedo, and the butler refused to admit him. According to a 1922 memoir:

> The story goes that Ward went out of sight, [and] pinned the skirts of his coat and the flaps of his waistcoat back far enough to imitate a dress suit. . . . Once admitted, he pressed into the circle of which . . . Livingstone was the center of attraction, and heard the great missionary lament that there was no-one alive who could give him information about the Niger [River] and the region [of Africa] which he intended to penetrate. To the surprise of the whole company, this roughly attired stranger . . . [replied]: "I think I can tell you, Sir, for I have just come from there."[2]

In 1860, Ward was appointed professor of natural sciences at the University of Rochester in upstate New York. Inspired by his mentor Agassiz, Ward created a superb teaching collection, and he went into business for himself. In 1873, Ward's Natural Science Establishment opened its doors in Rochester in order to provide schools and universities with specimens for the teaching of science. Like Robert Damon, Henry Ward personally collected many of the specimens he sold, he had agents all over the world, and he became an agent for Leopold Blaschka.

In 1878, Ward issued a 22-page catalog containing the Blaschkas' models of 630 different specimens. This does not mean that the Blaschkas stocked examples of every model they made. That would have been impractical. Instead, they required customers to order from their catalogs, paying in advance. Ward sent the orders to Dresden, and Leopold and Rudolf made the models. In 1888, when Ward published a new edition of his catalog, the number of models had increased to 700. The Blaschkas' ingenuity in devising new models, as well as the sheer quantity of models they made and sold to an astonished public, is staggering.

In 1886, George Lincoln Goodale, professor of botany at Harvard University, traveled to Germany to persuade Leopold to abandon his successful career of making models of marine invertebrates and to concentrate instead on producing models of plants. The result was the unique collection of botanical models, known collectively as the Glass Flowers of Harvard, which is described on pages 20–22. Many of the invertebrate models are spectacular, but the best of the botanical models are breathtaking.

2. Augustus Hopkins Strong, *Henry A. Ward: Reminiscence and Appreciation*, Rochester, New York: Rochester Historical Society, 1922, p. 28.

The Botanical Models (1886–1936)

Susan M. Rossi-Wilcox, Administrator for the Glass Flowers Collection and Curatorial Associate in the Botanical Museum–Harvard University Herbaria

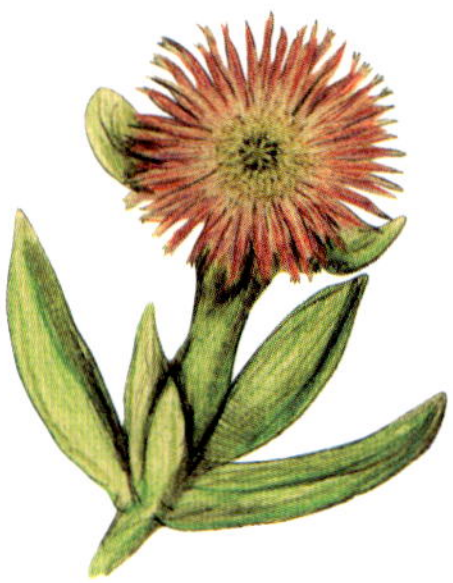

TROPICAL and temperate plants continuously bloom in their Victorian cherry wood vitrines as visitors to the Harvard Museum of Natural History marvel at their favorite flowers and the most noxious weeds. These nearly 4,300 botanical models represent roughly 840 species and 170 plant families in an incomparable collection popularly known as the Glass Flowers. Commissioned in 1886 by Prof. George Lincoln Goodale (Fig. 1), then director of the Harvard Botanic Garden, the exhibition was inspired by his desire to represent the full glory of the plant kingdom in the natural history museums that were then being developed at Harvard. This display would also help to supplement botanical courses because the numerous planted beds and greenhouses that Goodale oversaw at the Botanic Garden were subject to New England winters that limited the selection of teaching materials during the academic year.

Goodale searched for something that would be aesthetically pleasing, scientifically accurate, and much appreciated by both botany students and the general public. While peacocks can be stuffed and minerals can be meticulously polished, presenting plants as attractive displays is more problematic. Although lovely botanical models fashioned from wax-covered silk or papier-mâché were available as teaching aids, Goodale was impressed by the idea of using glass after he saw the zoological models in the Museum of Comparative Zoology at Harvard, which housed detailed glass models made by Leopold Blaschka and his son, Rudolf. This allowed members of Harvard's faculty to use replicas of organisms to easily point out morphological features during lectures. Goodale realized that the fragile zoological creatures were much like fresh fruits and flowers that quickly decay and could never hold up as an exhibition. He also knew that dried, pressed herbarium specimens and color plates from botanical texts had a limited appeal.

In 1886, Goodale traveled to Dresden, Germany, where the Blaschka family lived. Despite their initial resistance, Goodale ultimately persuaded the glass artists to accept a commission for a few plant models. The first shipment of these models was badly damaged by customs inspectors who hastily unpacked them, but the wafer-thin petals and leaves were convincingly lifelike. Goodale and his former Radcliffe College students, including Mary Lee Ware, believed that the Blaschkas' skilled hands and extensive botanical knowledge could create a most remarkable exhibition.

Leopold and Rudolf Blaschka agreed to a part-time, three-year contract. When it was time to renew the agreement in 1890, Goodale convinced Elizabeth Ware (Fig. 2) and her daughter, Mary (Fig. 3), who had become the benefactors of the Glass Flowers collection,

Acknowledgments. In this project, there are many colleagues to thank. Chief among them are Rosemarie Starlinger, who provided most of the transcriptions of the 1892 letters; Ulrich Jordis, who translated some of the 1892 letters; Hubertus von Posern, who translated the 1895 letters and other documents; and especially Laura E. Lukas, who translated the bulk of the 1892 transcriptions, including mine. All of the letters are housed in the Blaschka archives of the Botanical Museum–Harvard University Herbaria holdings. The notes on the drawings were translated by Erich Moraine, a flameworker, and Laura E. Lukas completed three other sets of notes. David E. Boufford and Gustavo A. Romero reviewed the captions for the drawings, and Richard W. Stratton provided a careful reading of the manuscript. Lisa DeCesare, Gustavo Romero, Sylvie Laborde, Jill Thomas-Clark, and Gail P. Bardhan helped to provide digital images.

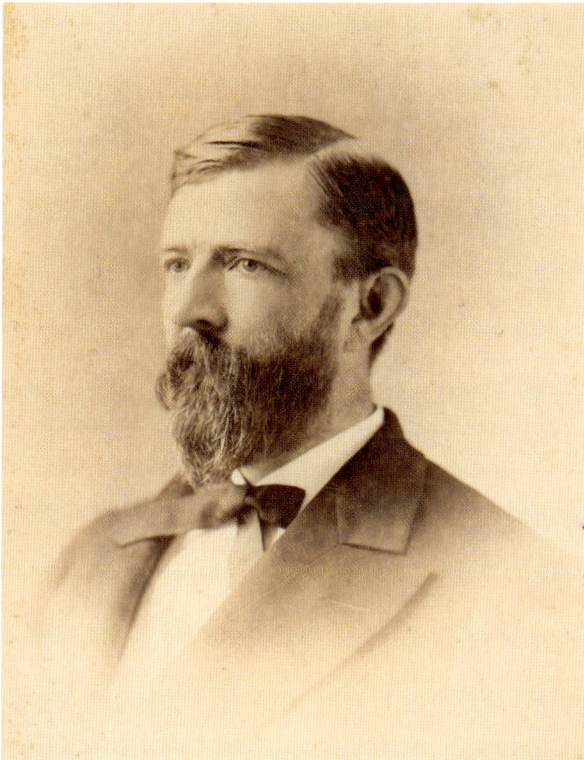

that a full-time, 10-year term would be more appropriate. The exhibition, which officially became a memorial to Charles Eliot Ware (Elizabeth's husband and Mary's father) on May 10, 1890, was titled The Ware Collection of Blaschka Glass Models of Plants. This collection represents the diversity of flora, with an emphasis on economically important plants that are used in everything from foods to medicines. The Blaschkas created three additional series of models, which show the life cycle of nonflowering plants such as ferns, mosses, and liv-

Figure 1
George Lincoln Goodale. From photographic album belonging to Leopold Blaschka. Rakow Research Library, The Corning Museum of Glass.

Figure 2
Elizabeth Ware. Botanical Museum–Harvard University Herbaria, Blaschka Archive, Cambridge, Massachusetts.

Figure 3
Mary Lee Ware. Photo dated 1907. Rakow Research Library, The Corning Museum of Glass.

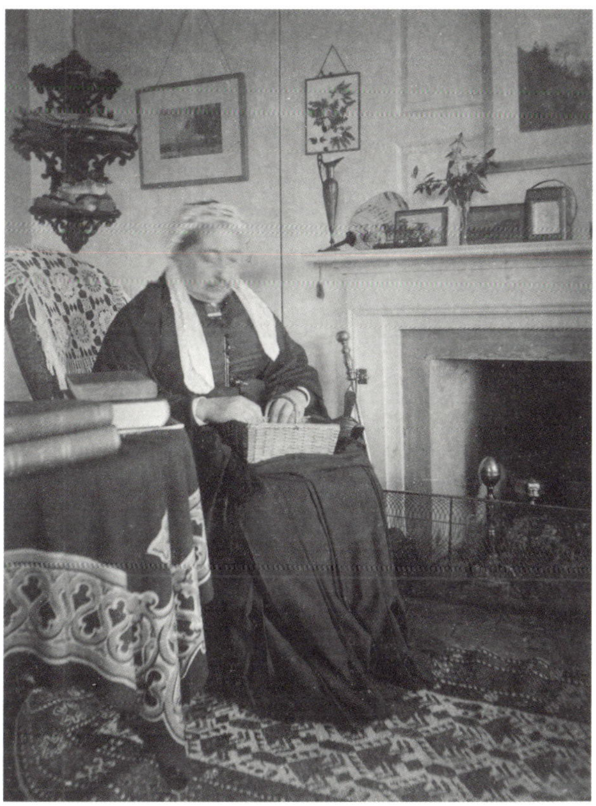

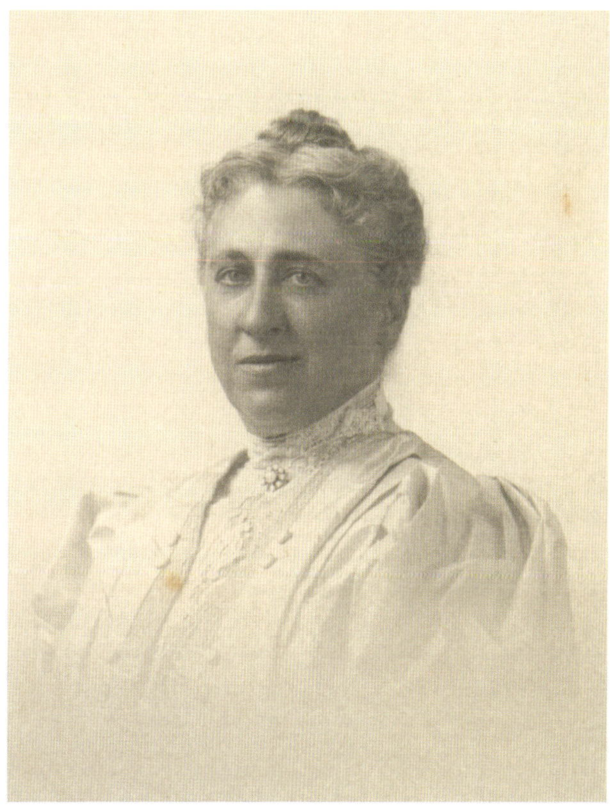

erworts; insects and pollination; and blighted apples, pears, and other fruits.

Visitors to the exhibition, particularly gardeners, often ask about the basis of the Blaschkas' models. Over the years, Goodale brought the artists a wide variety of seeds, cuttings, and plants from Cambridge. Since the climate of Saxony is similar to that of New England, the plants grew well. Although there are only a few contemporaneous written accounts of the Blaschkas' activities, 19th-century photographs of their home reveal the extent of the cultivation on their property. Reference specimens were also obtained from important gardens nearby, such as the Pilnitz Castle gardens and the Royal Dresden Botanic Garden, and from other state arboretums and botanical gardens in Germany. The Blaschkas' neighbor, the grand duchess of Mecklenburg-Strelitz, parted with a mature branch of her fig tree. Male and female flowering branches of the maidenhair tree (*Ginkgo biloba* L.) were obtained from the Grand Ducal Gardens in Karlsruhe (Baden).

By mid-1890, after they had made models of several hundred species to represent the major plant families, the Blaschkas had begun to exhaust their European sources. Although Rudolf was interested in undertaking a field expedition in North America, Goodale noted: "It seems worthwhile to induce young Mr. Blaschka to try his hand at reproduction of models from dry-specimens, exactly as the botanical artists do. He would have to be shown how this sort of work is done, and if the experiment is a success, there will be no limit to the material at command, for we can procure the dried specimens in every stage of development."[1]

This implies that the Blaschkas were used to, and preferred to have, living reference material. It also suggests the complications involved in adding exotic or difficult-to-cultivate species to the collection. Judging from some of the flat-looking models, relying on herbarium specimens must have worked for a while. But it later became evident that Rudolf's request would have to be considered seriously. This is indicated in a letter from Goodale to Elizabeth Ware on April 15, 1890, that outlined the conditions of the newly written contract. The Blaschkas offered to save enough money for Rudolf "to make a voyage to America in the winter of 1891–1892 in order to repair our broken models and make fresh studies." But Goodale told them "that when the time came for his voyage I would see that he had enough for his expenses. He is anxious to visit South America and study some of [the] plants of great interest in their native forests, which are known to Science only by poor drawings."

Colored pencils and drawing paper were carefully tucked into his luggage when Rudolf Blaschka left the cozy villa he shared with his parents in Hosterwitz, Germany (the family had moved there in 1888), to begin a journey to study the flora of the West Indies and North America. The northwest trip to Bremen on February 1, 1892, was pleasant until he reached Hanover, where the wagon "shook worse than a stagecoach on a rocky country road. . . . One could hardly balance oneself on one's seat." Yet, as the 35-year-old glass artist wrote to his parents later that day, he "found the experience extraordinarily nerve-strengthening and fun."[2] Upon arrival, he visited Ehrhorn, Emden & Mayer, shipping agents for the Blaschkas' models, who arranged travel documents and a first-class cabin on the steamship *Saale*. He sailed the next day from the northern coastal town of Bremerhaven.

Like his fellow passengers, Rudolf endured a southwesterly storm. It blew in as he was enjoying the fresh air on deck, and it continued while he was in the lounge practicing his English, and later when he was ensconced in a luxurious private room. "I took Father's advice and didn't participate in the six-o'clock supper," he recalled in a letter dated February 14, and he was among the few who fared well enough to emerge for breakfast. Ten days later, the ship docked in Hoboken, New Jersey, where William Francis Ganong, a botany instructor at Harvard College, arranged with customs officials for Rudolf's easy entry. In the same letter, Rudolf reported: "I met Mr. Ganong, a slim man of my height, about my age, and with a black mustache, who concerned himself about me in the most friendly manner and hired a hackney that brought us to the central

1. Undated but pre-1892 typewritten sheet by Goodale titled "Suggestions for Developing the Collection of Glass-Models."

2. All of the quotations attributed to Rudolf Blaschka in this account are excerpted from letters he wrote to his parents.

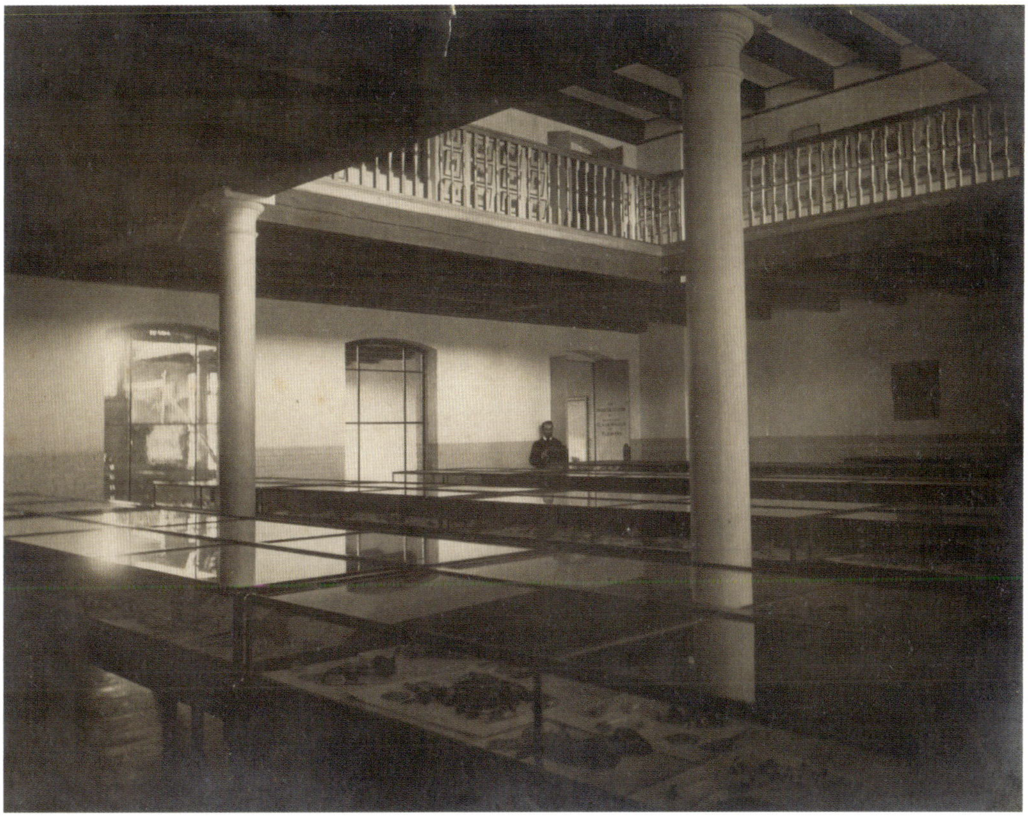

Figure 4. *Exhibit of glass flowers at Harvard Botanical Museum, 1892. Rudolf Blaschka is shown in the background. Rakow Research Library, The Corning Museum of Glass.*

station [Grand Central Depot] in New York City. The night train at 11 took us (in a sleeping car) to Boston, where we arrived at 7 in the morning."

Rudolf stayed with George Lincoln Goodale and his family in their Cambridge brownstone. They walked to the Botanical Museum, where, for the first time, Rudolf saw the dazzling exhibition of more than 250 sets of the models that he and his father had created, handsomely displayed in cherry wood vitrines (Fig. 4). He observed, in his letter of February 14, that "an iron staircase leads to the Ware Collection of Blaschka Glass Models on the second [American, third] floor. With delight and emotion, I saw all the old friends united here, and I can report to you with joy that everything, even the oldest things, are still nice in form and color, just as we delivered them."

He was equally excited about meeting his benefactors, Mrs. Elizabeth Ware, whom he described as "very strong and robust," and her daughter, Mary Lee Ware, "a large blond lady of very lively temperament," who regretted that she had been unable to visit the Blaschkas in Hosterwitz. Receiving the hospitality of Boston society and Cambridge academics alike, Rudolf quickly acclimated to life at Harvard. He attended dinner parties, as well as lectures at the Boston Natural History Society, and he viewed the orchid collection of Oakes Ames, who would become the second director of the Botanical Museum and continue the commission for Rudolf's production of glass models.

Rudolf began a diligent study of the flowering plants in the Harvard Botanic Garden, making reference drawings for himself and his father, which continued until the arrangements for fieldwork on Jamaica had been completed. Robert Cameron, the Scottish head gardener, accompanied Rudolf when they finally boarded the steamship *Bowden*, a banana boat owned by the Boston Fruit Company, for the six-day voyage to the West Indies. They disembarked at Port Antonio, on the north-

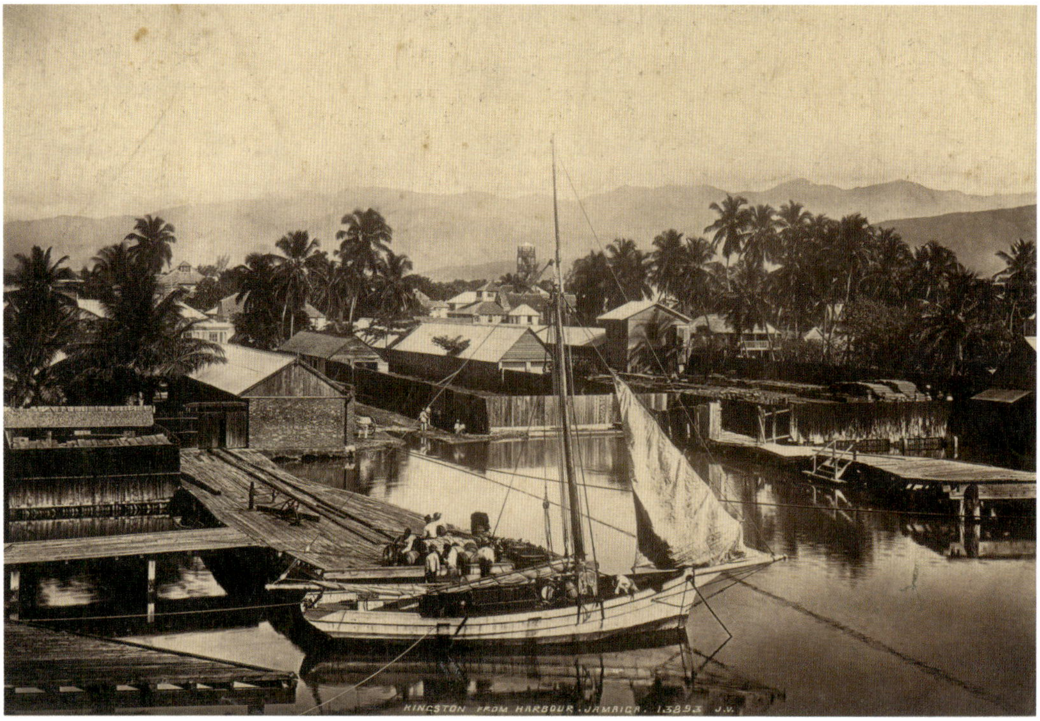

Figure 5. *Kingston, Jamaica, viewed from harbor, about 1891. Botanical Museum–Harvard University Herbaria, Blaschka Archive, Cambridge, Massachusetts.*

ern coast of Jamaica, on February 28, ready to collect the flora (Fig. 5).

The island's varied terrain, including sandy beaches, lush riverbanks, marshlands, forests, and mountains, offers ecosystems supporting tropical, subtropical, and temperate plant communities. The rich flora and fauna had been observed by Europeans from the time that Christopher Columbus returned to Spain with descriptions of the paradise's diversity. Jamaica sustains more than 3,000 flowering plants, of which a third are endemic. These include hundreds of orchids, which were nurtured by early botanical gardens. In 1862, Castleton Gardens (Fig. 6), located in the parish of Saint Mary, was planted with material that had been rescued from the frequently flooded Bath Botanical Garden in Saint Thomas, the second-oldest garden in the Western Hemisphere. Rudolf studied these species, particularly those of the *Croton*, *Bougainvillaea*, and *Jacaranda* (Figs. 7 and 8). In 1892, Castleton Gardens was being developed into one of the most significant collections on the island.

Hope Gardens in Kingston, Jamaica, which was renamed the Royal Botanical Garden in the 1950s, was founded as a public garden on the Hope family estate. It had been established only two decades before Rudolf Blaschka's visit. In a letter dated March 6, he related:

> Hope Garden [*sic*], where I am now staying, is still newly planted, as just four years ago thick forest was in this place, but there are still many old trees and useful plants that greatly interest me. So I could study many flowers and fruits of tropical bananas, for example, the coco tree, mango (*Mangifer indica*), the star apple (*Chrysophyllum*), and both of the coffee varieties [that] are grown in Jamaica, *Coffea arabica* and *liberica*. Today I had the pleasure of seeing the flowers of the cinnamon tree, *Cinnamomum officinale* [see drawing on page 103], which grow almost wild here. It is very interesting to see such plants; for us at home, in the garden or in the greenhouse, they grow to be only small trees or bushes, while here they are massive.

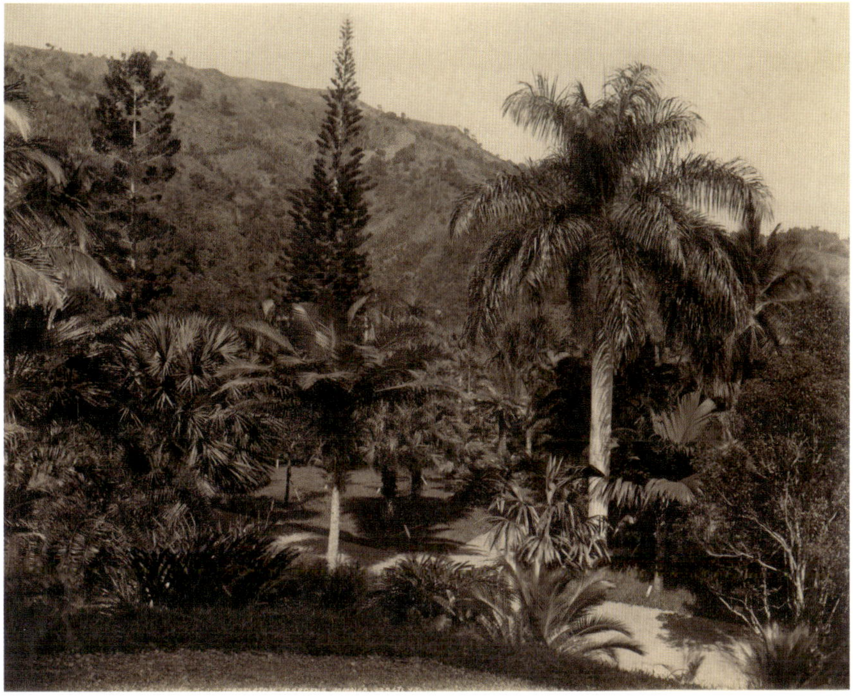

Figure 6
Castleton Gardens, Jamaica, about 1891. Botanical Museum–Harvard University Herbaria, Blaschka Archive, Cambridge, Massachusetts.

Figure 7
Jacaranda filicifolia *D. Don from Castleton Gardens, Jamaica. Drawing by Rudolf Blaschka, 1892. Rakow Research Library, The Corning Museum of Glass.*

Figure 8
The Blaschkas' model of Jacaranda filicifolia *D. Don, based on the drawing shown in Figure 7. Photo by Hillel Burger, courtesy of Botanical Museum–Harvard University Herbaria, Cambridge, Massachusetts.*

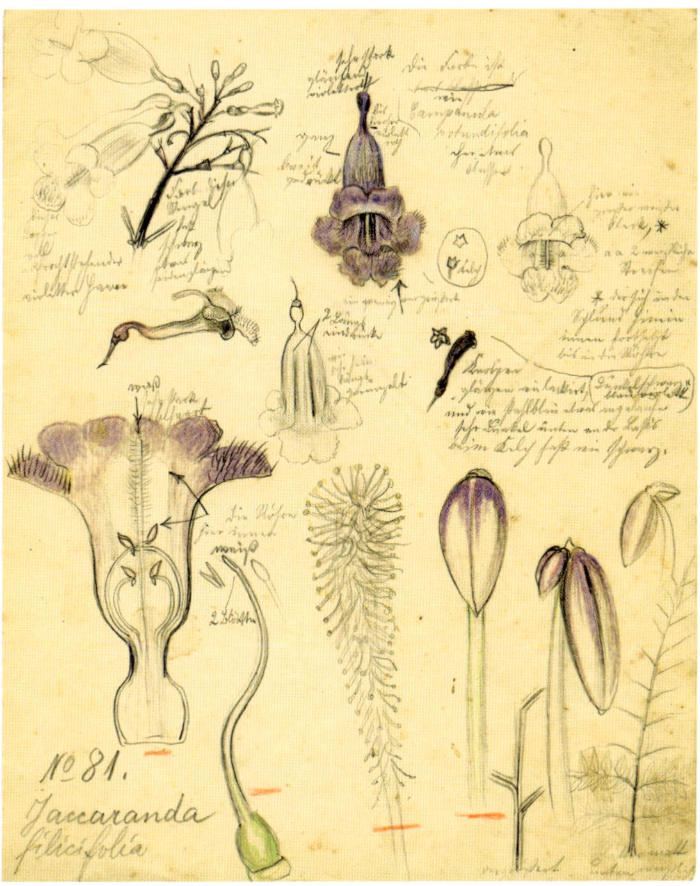

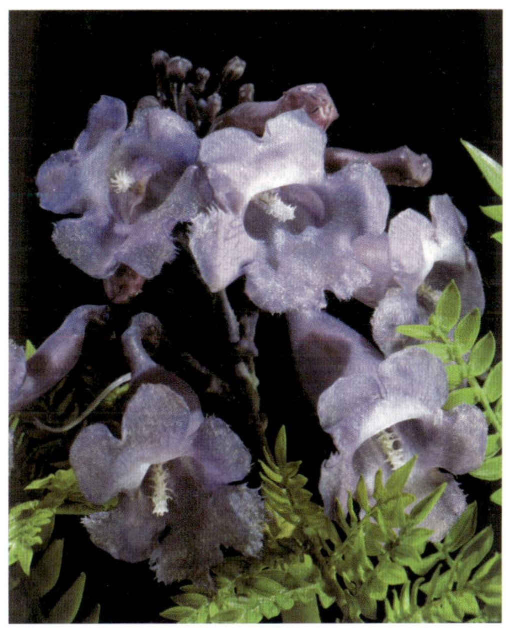

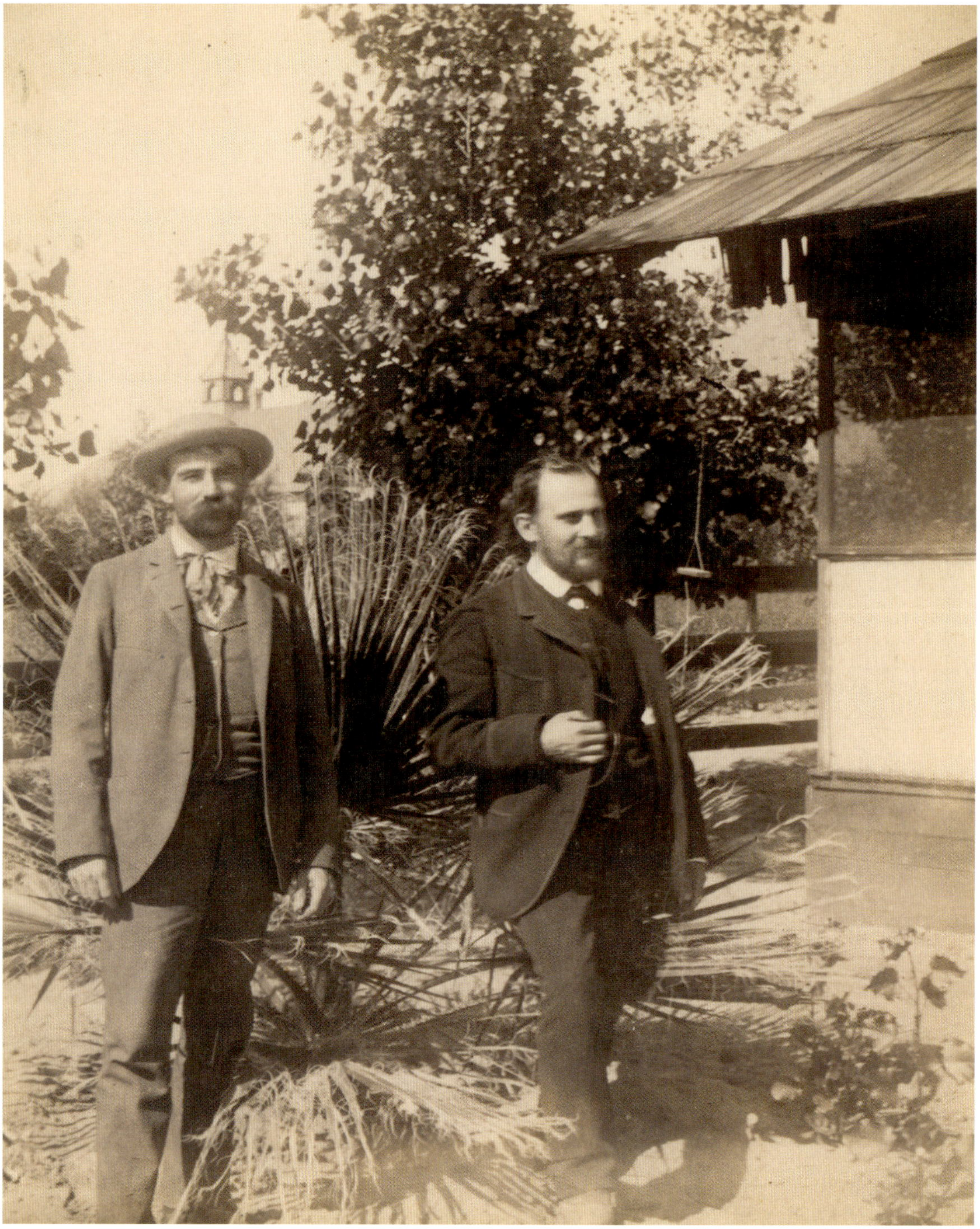

Figure 9. *William F. Ganong (left) and Rudolf Blaschka, 1892 field season. Photo dated April 28, 1892. Rakow Research Library, The Corning Museum of Glass.*

On the island, Rudolf and Cameron were occupied with such practical matters as transporting the microscope, drawing supplies, glass specimen jars, collecting equipment, and luggage, either by sea or by land, from the various sites. Rudolf took these challenges in stride. From a banana wagon pulled by donkeys, he viewed what he described (in a letter of March 6) as "the romantic, wonderful, beautiful mountain valleys and . . . slopes [that] will remain unforgettable for me" on the way to Castleton Gardens. He left "the pleasure of exploring the mountains, and the terrible hot sun for collecting plants, to Mr. Cameron."

Their routine was soon established. After Rudolf had completed his drawings, Cameron pressed and dried the herbarium specimens and pickled any extra flowers or fruits in the local white rum, diluted with water. In his "airy room" at his "quiet work, just like at home," Rudolf produced about five drawings a day. He ultimately made detailed observations of 106 tropical plants from Hope Gardens, Castleton Gardens, Port Antonio, Spanish Town, and other areas during his month-long stay on Jamaica. "It is most important that I study these things here in the tropics because they are mostly trees and shrubs that would not bloom in greenhouses," he wrote on March 21.

Although Rudolf had "a very small camera" on loan from the Botanical Museum, he regarded it as a waste of his time. "Photographs are totally unnecessary to our studies," he reported on March 23, "since I would much rather draw than toil with boring photography." Goodale had attempted to teach Cameron how to document the plants and landscapes, but the gardener ignored this task.

When the tropical representatives of important economic plants had been documented, the Botanical Museum's mission had been served, but life in Jamaica was not all work. Rudolf socialized with his hosts, played the piano at gatherings, and found, on those balmy evenings, that "it was also pleasant to sit on the veranda, which was surrounded by the wonderful blossoms of *Datura arbora*" (letter of March 21).

Many aspects of island life fascinated Rudolf, but the banana industry held a special attraction—or at least it passed the time during the six-day return voyage on a steamship that was loaded with the green harvest. Rudolf had already acquired a taste for bananas, which were shipped twice a week to Boston distributors, and Goodale ate bananas daily with his lunch. Prompted to try them, Rudolf acknowledged, on February 19, that "they taste good, and I like eating them." Catching the spirit of economic botany, he shared another experience with his parents on April 10 by mailing samples of "the best sort of Jamaican coffee, *Coffea liberica*, from the plant, and I'm sending you the ones I picked so dear Mother can make real Jamaican coffee." Although the drinks, foods, and spices of Jamaican cuisine must have fascinated him during his visit, he rarely commented on them, preferring to discuss the models of tropical comestibles, including bananas and coffee.

Upon his return to Massachusetts, Rudolf allayed his parents' fears about the yellow fever epidemic that had been reported on Jamaica. He also assured them that his letters from Cambridge would reach Germany in only 12 days, instead of the 21 to 23 days required for communications that had been sent from Jamaica. Based in Cambridge from April 5, Rudolf spent the week drawing species of elms, American hazelnut, willows, and other plants at the Arnold Arboretum and the Botanic Garden. Entertainment was never lacking. He attended a circus performance with William Ganong, ascended the Bunker Hill Monument, saw friends at a Boston Natural History Society meeting, and dined with various families.

Rudolf's delay in setting out immediately for the West was resolved when Ganong agreed to accompany him as his field partner (Fig. 9). In many ways, this Canadian botanist was a good choice, for he spoke German and was interested in plant communities. He would later make significant scientific contributions to the newly emerging study of ecology. Although they had $600 at hand and about a $3 per diem for food, Ganong believed that they could reduce expenses by boarding with individual families. They planned to travel by train directly to Chicago and then through Missouri, Kansas, Colorado, and New Mexico before stopping to collect specimens in Arizona and southern California. After completing about two months of fieldwork, they would return by way of San Francisco and Colorado on the Northern Pacific train.

Finally, on April 12, they set out, dressed in lightweight three-piece wool suits, as was the custom. The next day, the avuncular Goodale wrote to Mrs. Ware, who had financed the fieldwork: "Rudolf started last night at 6:45 with Mr. Ganong. They went off in good spirits and promised to let us know frequently about their movements. I shall, of course, keep you informed."

In a letter of April 18, Rudolf reported that "it made me very happy to travel through the unbelievable stretches of land of North America to see the different interesting cities and landscapes. . . . The coast of Lake Erie . . . offered many interesting views and landscapes while the vegetation was still in its winter state." He and Ganong spent a day in Chicago exploring by "streetcars and the elevated line in every direction and looked at the buildings of the world's fair, the Art Museum, the skyscrapers," and other noteworthy attractions.

Having seen the metropolis, they boarded a comfortable first-class sleeping car and left Chicago behind. "The plains of Kansas seemed endless," Rudolf wrote in the same letter, adding that on "the wild prairie in Colorado . . . we could go for hours without seeing a house. As far as the eye can see, there is nothing but plains full of dry grasses, which immediately catch fire from the locomotive and sometimes keep burning along the train." Then came the "amazingly beautiful" trip through the Rocky Mountains, down to the savannas, and finally into New Mexico, "where herds of thousands of animals [buffaloes] are grazing." The grasslands blended into "hundreds of miles where nothing grows but one species of yucca, cacti, and desert plants."

On Easter Sunday, they reached the Rio Grande, which was then "the border between the United States and the Republic of Mexico." They explored the border town of El Paso, Texas, and took a tramway into Mexico, but the vegetation was unpromising and they moved on.

After a total of six nights spent on the train, Rudolf described himself "as used to the shaking as an old sailor." Fortunately, they reached Tempe, by way of Maricopa, Arizona, during the few weeks of spring when the desert plants are in full bloom. The English hotelier and his Bavarian wife offered "nice rooms" for $1.50, food at $1 a head, and conversation in German. Rudolf and Ganong planned "to stay a few days because there's a lot of good study material" (letter of April 18). They took group excursions or rented wagons, riding out into the chaparral early in the morning, but Rudolf usually lingered in order to draw, and he informed his parents on April 22 that he had created 25 studies in the Tempe area.

As they crossed the California border en route to San Diego, Rudolf remarked on the huge desert, "where nothing grows on the cracked sand and rocky ground." The landscape changed as the huge yucca species and blooming opuntias gave way to the vibrant plantings, agricultural fields, and lush orchards that surrounded the city (Fig. 10). In San Diego, Rudolf collected less frequently and spent more time in the hotel, sketching what had been brought to him from distant sites. He did travel to Tijuana, Coronado, Hesperia, and San Ber-

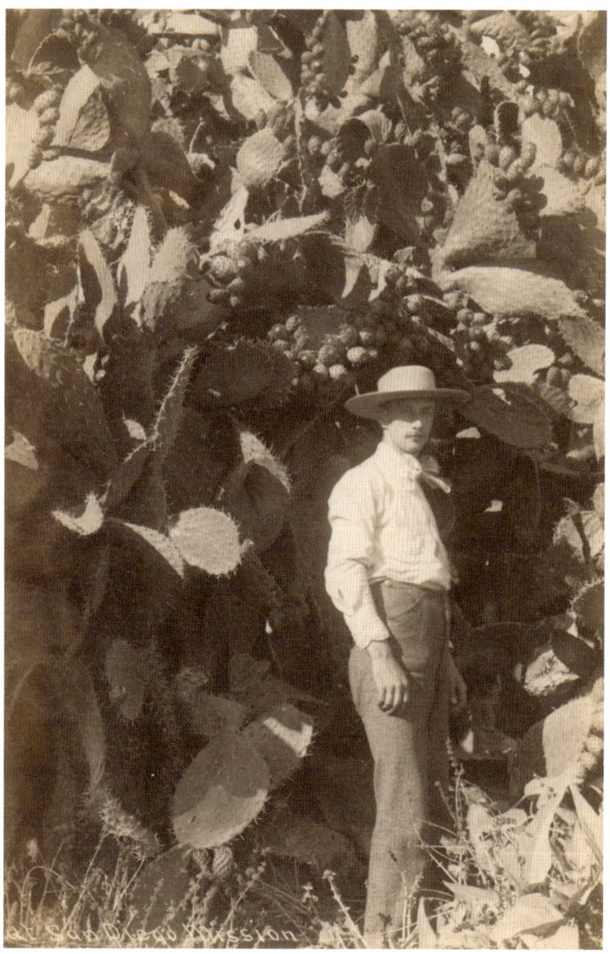

Figure 10
San Diego Mission, to which Rudolf Blaschka traveled. The man shown in the photograph is not identified. Rakow Research Library, The Corning Museum of Glass.

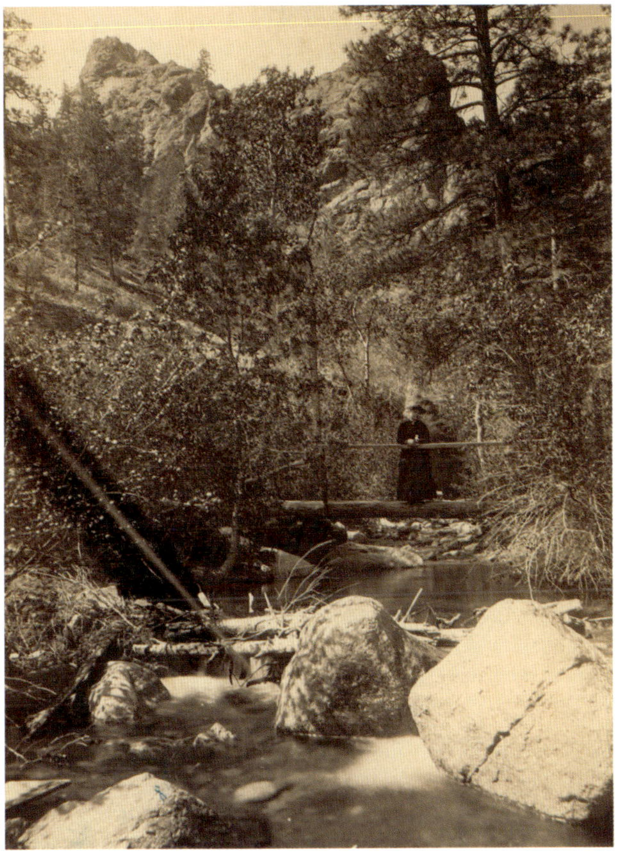

Figure 11
South Cheyenne Canyon, photographed by W. E. Hook, Colorado Springs. Rakow Research Library, The Corning Museum of Glass.

Figure 12
*Cacti (*Opuntia bigelovii*) near Coyote Hills, Colorado desert, 1890. Rakow Research Library, The Corning Museum of Glass.*

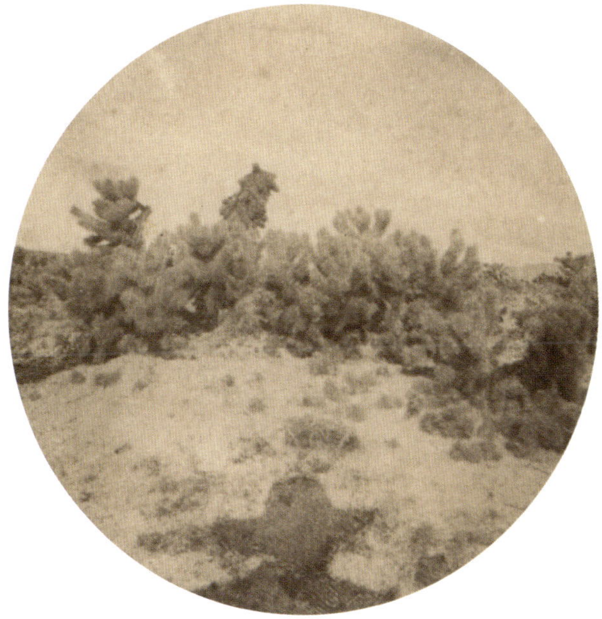

nardino to observe the forests of conifers. Agua Caliente (now Palm Springs), their next stop, was considered a boom town that had entered a period of decline, but this remote area offered new plant species, a hot-water spa, and a comfortable hotel owned by a physician-botanist, Dr. Murry.

Rudolf's letters home were filled with accounts of his new experiences. On May 15, he wrote, "This scientific gypsy life is fun, but I will be very glad to exchange it for the quiet life I'm accustomed to." Ganong reported to Goodale that his field partner was an industrious but homesick traveler who was unlikely to welcome the Southeastern trip that was being planned to follow their return to Cambridge. With the flurry of work and the frequent change in diet, climate, and conditions, Rudolf became ill in Oakland, and he was on the mend for the next two weeks. He later stated that he had been diagnosed with malaria, and an injury from a cactus spine certainly did not help his disposition. In addition, he complained that he was being treated poorly, money was being spent too liberally, and he was often unable to see plants in situ as he had wished.

But on they went, collecting near Colton, Fresno, Oakland, and San Francisco. Rudolf's letter of May 15 reported that the bright orange-red California poppies, *Eschscholzia californica* Cham., were "so common and grow so thickly that the mountains glow orange," and that the ubiquitous clover, *Orthocarpus purpurascens* Benth., provided "a soft pink color in great masses." Outside Santa Cruz, they visited the sequoia forests, concluding their productive tour of California.

They then headed eastward, and by May 22, they arrived in Colorado Springs. Two days later, Rudolf informed his parents:

> The botanical results were surprisingly rich. Through the gorge, between rock formations, there rushed a small river surrounded by various vegetation [Fig. 11], where we found a great many things in bloom that were much prettier flowers than those in the desert [Fig. 12]. Upon arrival at home [his hotel] yesterday,

> I quickly drew two more species and today five. Tomorrow I will have to study again because today we found several new things. I am now working in a very casual manner in order not to tax myself too much. . . . Don't think that if only I could work more quickly we could collect much more; at this time of year, there isn't a large selection. In addition to that, the people in Cambridge are picky; they don't want this or that species because it's already represented, etc. In order to complete 600 to 800 species, I would have to spend the whole year wandering around America, and I really don't want to do that. Until now, I have drawn and studied 200 species, and I think that, when I leave America, I will have collected 400. At any rate, we found so much good material here that we decided to stay until Saturday.

The field partners endured a 40-hour train trip from Colorado to St. Louis, Missouri, where the largest botanical garden in America had been established in 1856. They had already met its director, William Trelease, when he was in San Diego. Having access to the Missouri Botanic Garden saved time in the field because the plants were close at hand and already identified. Even *Darlingtonia*, the elusive pitcher plant that had been requested by Goodale, for which they had searched the Shasta Mountains, was in bloom here. Five days later, they were back on the road again, with stops in Chicago, Buffalo, and several locations in Ontario, Canada, including Clinton and Niagara Falls. On June 7, they arrived in Cambridge, where Goodale reported to the Wares that "a more homesick body than R.B. was never seen in the world. He has decided to repair the unmended breaks, and to see a seedsman in Southwick [Massachusetts], but he has solemnly promised his father to be back in Germany by 'Gooseberry time,' that is, the last of June. He is looking very well and is anxious to get at work at the models again."

On June 10, Goodale informed the Wares that, with Rudolf's collections in Jamaica and the United States, he believed that he had enough specimens to keep him and his father busy for at least three years. Goodale also thought that they could spend another three years on the seeds and cuttings that Rudolf had brought back with him on June 17. Although the surviving archive of specimens, drawings, and labels is incomplete, Rudolf probably documented upwards of 350 species, and he and his father subsequently created more than 250 sets of models based on this 1892 field season.

The work proved so valuable that Rudolf returned for a second field season three years later. Capitalizing on his 1892 plan to collect in the Southeast, Goodale suggested that this trip culminate in Mexico. Rudolf, who was by now a confident traveler, took a train to Bremen, purchased his ticket on the *Fulda*, and departed from Bremerhaven on June 9. After battling rough seas, the ocean liner arrived in Hoboken, New Jersey, on June 18. When the frugal Rudolf learned the price for hiring a carriage to New York's Grand Central Depot, he sent his extra bags ahead to the Botanical Museum. He then took a ferry across the Hudson River and a streetcar to the train station. The 11 a.m. express train delivered him in Boston six and a half hours later. He decided on the Thorndike Hotel, booked a room at $1 a night, and arrived punctually at the Goodales' home in Cambridge for dinner at 8 p.m.

The problems that had been experienced in 1892 were discussed openly and resolved. Rudolf informed his parents on June 19 that Goodale had retained, as an assistant, "a young man, a poor student named [Joseph William] Blankinship, who traveled extensively with great success in plant collecting." After Rudolf had made a few studies at the Arnold Arboretum in Jamaica Plain, Massachusetts, a quiet room in the Botanical Museum was reserved for him so that he could draw the rare mountain plants that had just arrived from New Hampshire. Within a few days, he and Blankinship, who was only five years younger, were off to the coastal marshes near Plymouth, Massachusetts, where Rudolf acknowledged that he had a worthy field companion. When they returned to Cambridge, Rudolf relocated to an apartment and ate with the students at the Foxcroft Club near the museum. His mission now was to study all of the flowering species in the vicinity before he and Blankinship headed south. In Concord, they found water plants in bloom. On July 1, Rudolf wrote: "Mr. Blankinship undressed without any hesitation and pulled out these plants, although the water reached his hips. If I had been alone, I never would have done this."

Ten days later, Rudolf and Blankinship began their trip south, although their exact itinerary was still uncertain. They planned to take a train to Washington,

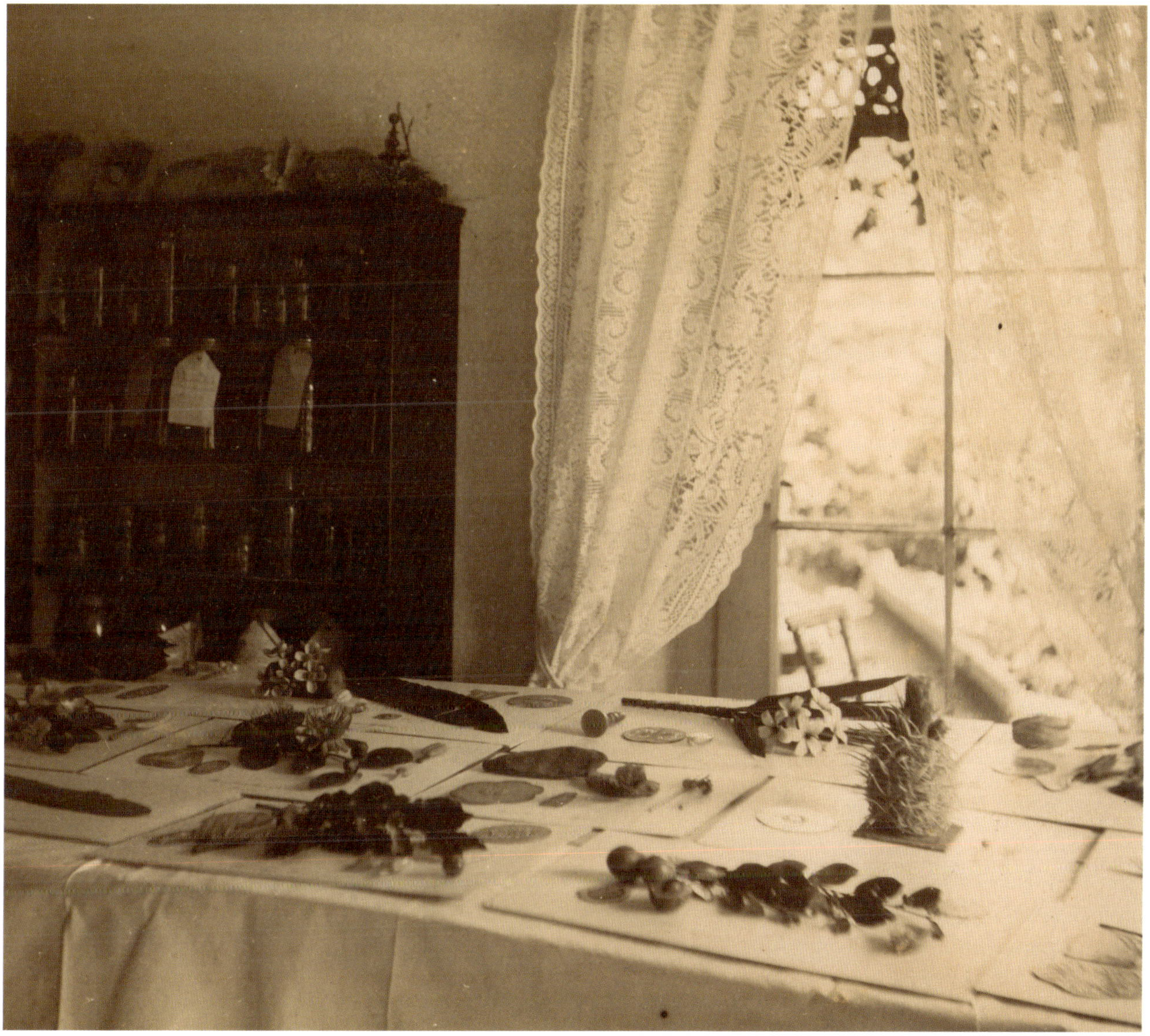

Figure 13. *Models of glass flowers in the Blaschkas' studio before they were shipped to the Harvard Botanical Museum. Probably photographed by Rudolf Blaschka. Photo dated October 1891. Rakow Research Library, The Corning Museum of Glass.*

D.C., for a sightseeing tour before boarding a steamer down the Potomac River through the Chesapeake Bay to the harbor at Norfolk, Virginia. The numbers of flowering plants they found in each place they visited would determine the length of their stay. Wilmington, North Carolina, was the next stop before they collected indigenous plants in the mountains. Afterward, they planned to go to Chattanooga, Tennessee, and possibly through parts of Georgia and Alabama, before heading west across Mississippi, Arkansas, Missouri, and the Indian Territory of Oklahoma. A train from Texas would then take them to Mexico City. Cyrus G. Pringle, a Vermont botanist who specialized in the rich Mexican flora, would join them in August for a few weeks.

The delay in receiving letters from Germany, which often had to be forwarded to remote locations, is reflected in the fact that Rudolf was utterly unaware that his father had suffered a stroke on June 27, followed by a second, fatal one on July 3. His chatty letters focused on chaotic Fourth of July celebrations, excursions to nearby towns, and the completion of his 61st drawing. In his closings, he always noted that he had not received any letters and was longing for news from home.

The uncertainty of his movements made it difficult to redirect his mail. The physician's letter concerning the circumstances of Leopold's death waited at the Botanical Museum, along with a heartbreaking letter from Mrs. Blaschka that had been written five days later. In that communication, dated July 8, she disclosed her husband's last wish: "In his delirium, his intention was to go to New York because there it would be much nicer than Hosterwitz, his Rudolf would be with him, and he could work together with him." While she wanted her son to return as soon as possible, she also reassured him that the priest had given a worthy eulogy at the funeral, and now close friends, her sister, and her maid were taking good care of her. Meanwhile, the travelers continued south, not staying long enough in any one place to telegraph their address to the museum so that letters could be forwarded. Clearly the fieldwork was going well. Rudolf's letters home continued with exuberant descriptions of new sights and progress. By the time he left Virginia, he had finished his 70th sketch. On July 15, he wrote: "I wish the illustrating and studying would go much faster, but often I find totally strange species of plants in my hands that require a thorough study. Mostly I think it is good to investigate everything locally instead of [making] hasty sketches. Because those are very hard to read later, it is my intention to finish as much as possible."

When he reached Chimney Rock, North Carolina, the letters could finally be rerouted, and on July 23 he learned the devastating news of his father's death, nearly three weeks after the fact. He immediately returned to Cambridge, leaving Blankinship to collect plants as planned. Goodale was supportive and helpful. They renegotiated the contract for the models as a one-man operation at twice the price. Goodale secured a parlor car heading for New York City, and he saw Rudolf and his reference materials aboard. The next day, Rudolf boarded the steamship *EMS*, and he arrived home on August 9. He then worked in his studio, sometimes with the assistance of his mother, to complete the models his father had begun (Fig. 13). At that point, roughly 70 percent of the collection had been jointly made by Leopold and Rudolf. Rudolf then created 62 sets of models from the 1895 reference specimens.

From 1896 through 1936, when the last shipments of models were received, Rudolf added more than 200 species, including several series on grasses, insect pollination, and the progression of fruit blight. The contracts, which were renewed while Goodale oversaw the project, were resumed after World War I, and they continued in 1923 with Oakes Ames, who became the next director of the Botanical Museum.

Mary Lee Ware remained an understanding benefactor, which allowed Rudolf to take the time he needed for glassmaking experiments to develop his own colored glasses. There were years in which he did not send any models, but he kept her, Goodale, and, later, Ames informed of his undertakings. Mary encouraged the artist, and she always came to his aid financially. She also visited him and his mother in 1928, and she reported his progress to Ames on October 3. She said of the 71-year-old Rudolf: "I was daunted to see what seemed a little old man, legs that were not strong, very rounded, stooped shoulders and an exceedingly white face. He must have dropped nearly two inches in height, his hands were somewhat out of shape from rheumatism." Mary realized that he was anxious for her to see the new work because it had been six years since the last shipment. After he had been reassured of her satisfaction, Rudolf showed her his new techniques for enameling the models with his powdered glasses. "His movements are quiet, deft, soft in laying down or taking up where speed or a miscalculated movement might ruin the work of hours, . . ." she wrote. "It all leaves you breathless that anyone can and will do such work. . . . Mr. Blaschka's head and bearing are very expressive, and I wished I could catch a photograph of his profile as he stood for a few moments, a plaque with a model on it held in both hands. His whole expression of absorbed, concentrated study was worth keeping, had it been possible."

Rudolf continued to work with that same devotion and fascination on his last series of models: rotting fruits and fungi. Unfinished models remained on his lampworking table when he died on May 1, 1939. With loving reverence for those models, his wife packed and stored the studio materials until they were purchased decades later.

About the Drawings

THE FOLLOWING SECTION of this book contains 57 drawings, 27 (and three prints) of invertebrate animals and 30 of botanical specimens, that Leopold and Rudolf Blaschka used in constructing their glass models. The botanical drawings and many of the drawings of invertebrates are quite detailed.

The preparatory drawings for models of invertebrates were created from several sources: illustrations in scientific publications, specimens preserved in alcohol, and live animals kept in aquariums. Unlike the botanical drawings, which were made on well-documented expeditions into the field, most of the drawings of invertebrates are difficult to date. When they were copied from printed sources, we know that they cannot have been made before the original publications, whose dates are known. In other cases, it may be reasonable to conclude that the drawing in hand predates the first documented occurrence of the model it depicts, although we cannot be certain that no earlier drawing of that model existed. We hope that, by constructing a concordance of the drawings, the printed sources, and the dates at which individual models were acquired, we may eventually construct a chronology for the Blaschkas' drawings of invertebrate animals. This is one of the many ongoing projects connected with the Blaschkas archive in the Rakow Research Library of The Corning Museum of Glass and with models of invertebrates in natural history collections worldwide.

The taxonomic classifications of the invertebrates shown here accord with those used by the Blaschkas on their drawings and in their catalogs, although, in many cases, subsequent research has led to changes in nomenclature. In this publication, the drawings are ordered from the simplest organisms to the more complex forms.

Leopold's earliest models were copied from *Actinologia Britannica: A History of the British Sea-Anemones and Corals*, written by the British naturalist Philip Henry Gosse. Several of the drawings of invertebrates in this book are printed pages from Gosse's volume. They show just how accurately Leopold replicated, in a three-dimensional form, the shapes and details found in Gosse's illustrations. Some of the illustrations and their corresponding models are rather crudely rendered, and the colors are garish. At this point in his modelmaking career, Leopold was able to work only from drawings in zoological textbooks, and the quality of his work was, to a large degree, dictated by that of the illustrations he copied.

The working drawings that Leopold made from *Actinologia Britannica* reflect two distinct influences from Gosse's research. First, Gosse showed the animals in their natural habitats, and Leopold's models followed his example. Second, Gosse's illustrations portray animals on rocks in underwater scenes, and this detail, copied by Leopold in his drawings, may have prompted him to mount his models on decorated plaster bases.

The botanical drawings are quite different. These plants were carefully drawn in their natural settings by Rudolf Blaschka, who then proceeded to dissect them in order to highlight their various parts, either life-size or magnified, as details that accompanied the main drawings. The 30 examples represent equally the tropical and temperate plants that Rudolf assiduously studied. They include aquatics, herbs, vines, shrubs, trees, and a fern. Nineteen of the drawings are from the 1892 field season, and 11 are of specimens collected in 1895, which accounts for the nonconsecutive and duplicate drawing numbers. All of these drawings were made into models for the Harvard collection.

The detailed illustrations provide reference information on color and dimensionality that the herbarium specimens lost as the plants were pressed and dried. Rudolf's studies often include top, front, back, and side views of flowers and internal structures such as stamens, which produce pollen, and pistils, which are fertilized by the pollen to develop into a seed-bearing fruit. When these structures are fused or have a more complex morphology, he drew cutaway views to record the size ratios and the placement of each part of the plant. Sometimes a small diagram is provided to show the relationship of floral parts (see, for example, the drawing on page 107). Transverse and longitudinal sections of a fruit are often shown on the same page, although many of the studies have a second page with more details about the fruit and seeds.

Rudolf wrote in *Kurrentschrift*, an old form of German handwriting that is distinctly different from modern scripts.

Rudolf's notes contain critical information about the plants. Sometimes he provides numbers denoting color that probably correlate with the dry pigments that he and his father used to paint parts of the colorless glass models. In a letter to Leopold about repairing some models, he asked for "our regular dry paints from Weingärtner, etc." and "permanent green from Müller and Hennig, Prague Street."[1] These two suppliers and others have been identified in Blaschka studio materials purchased by the Botanical Museum and The Corning Museum of Glass, where they reside in the archives. The numbers recorded on the drawings are 1–4 for shades of white, 7 and 8 for yellows, 11–14 for oranges from yellowish to flaming red, 19–26 for a range of greens, and 29 and 30 for shades of violet. The colors are often given with other descriptors, such as shiny or matte textures, opacity, and the gradation of color change along the structure. Rudolf commonly employed familiar terms, such as "almost glasslike in transparency," and textile references such as "shiny like royal silk" for waxy surfaces and "wooly brown" to refer to hairiness. Twists, furrows, creases, wrinkles, stripes, and spots are all designated. At times, the size of an enlargement is provided. Some illustrations show the progressive opening of the bud to a fully mature flower with pollen-laden anthers or a receptive stigma. Any unusual features of the genus or species were captured in additional sketches or exploded views that Rudolf and his father would find helpful as they consulted the drawings in order to create three-dimensional models.

As a biologist, Rudolf Blaschka continued to draw what he observed in his own backyard well after he had completed his field seasons in the New World. Many of his later studies were used for models illustrating grasses and an insect pollination series in which four-inch "humble bees" hover above mature flowers. Collectively, the illustrations and herbarium specimens document the Glass Flowers collection at Harvard and verify claims of scientific accuracy. They offer invaluable insights into modelmaking from the perspective of the history of science, but more significantly, they reveal the satisfaction that Rudolf received from creating his technically accomplished studies. In a letter to Walter Deane, an American colleague, on October 23, 1899 (written in English), he remarked:

> I think I belong to that same order of men as you, to the true lovers of nature. On every walk I take, there must be something to study of nature, it may be a plant or insect or bird or whatever. I think a man can never finish these studies and is never too old to learn from nature. Already when I was a child I had passionately longed after an opportunity for visiting foreign countries on purpose to study and collect the strange products of nature there. This wish has been strangely fulfilled by the glass-flower engagement. What I saw and learned from nature on those trips in America gives me very sweet hours of remembering for all life.

1 Undated letter, written between February 13 and 19, 1892. Both of these businesses were local. Brückman and Weingärtner, Farbewaaren-Handlung in Dresden, offered a variety of chemicals. Müller and Hennig supplied glazes (*fein geriebe Porzellanfarbe*). Only one package had a number, F45 ("Chamois"), for a buff- or tan-colored glaze.

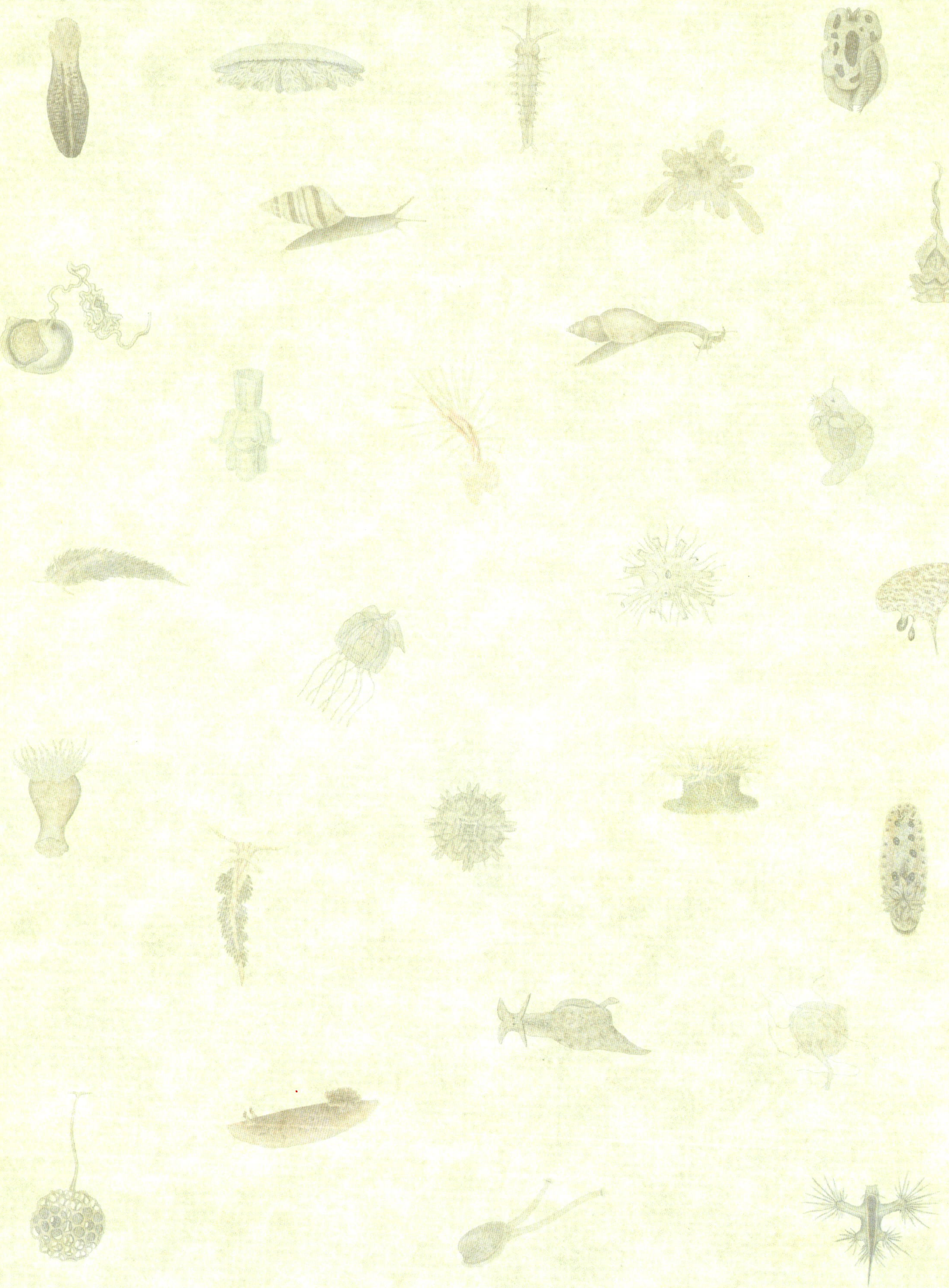

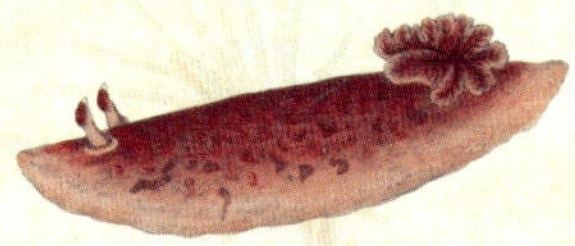

Drawings of Invertebrate Animals

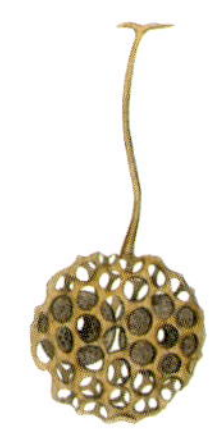

1.

Drawings 1–3:
Raphidiophrys elegans
Hertwig and Lesser
Model 636
Drawings 4 and 5:
Clathrulina elegans
Cienkowski
Model 635
Radiolarians
Order: Radiolaria
Ink and watercolor on paper
41 cm x 32 cm

Radiolarians are microscopic protozoans found in the upper layers of the world's oceans. They have skeletons, usually composed of silica, that are characterized by their intricacy and symmetry. The skeletons of dead radiolarians settle on the seabed, forming layers that, after millions of years, may be transformed into sedimentary rock.

Annotations record that drawings 1 and 2 show the animals magnified 500 times, while drawings 3 and 4 (and presumably 5) are magnified 1,000 times.

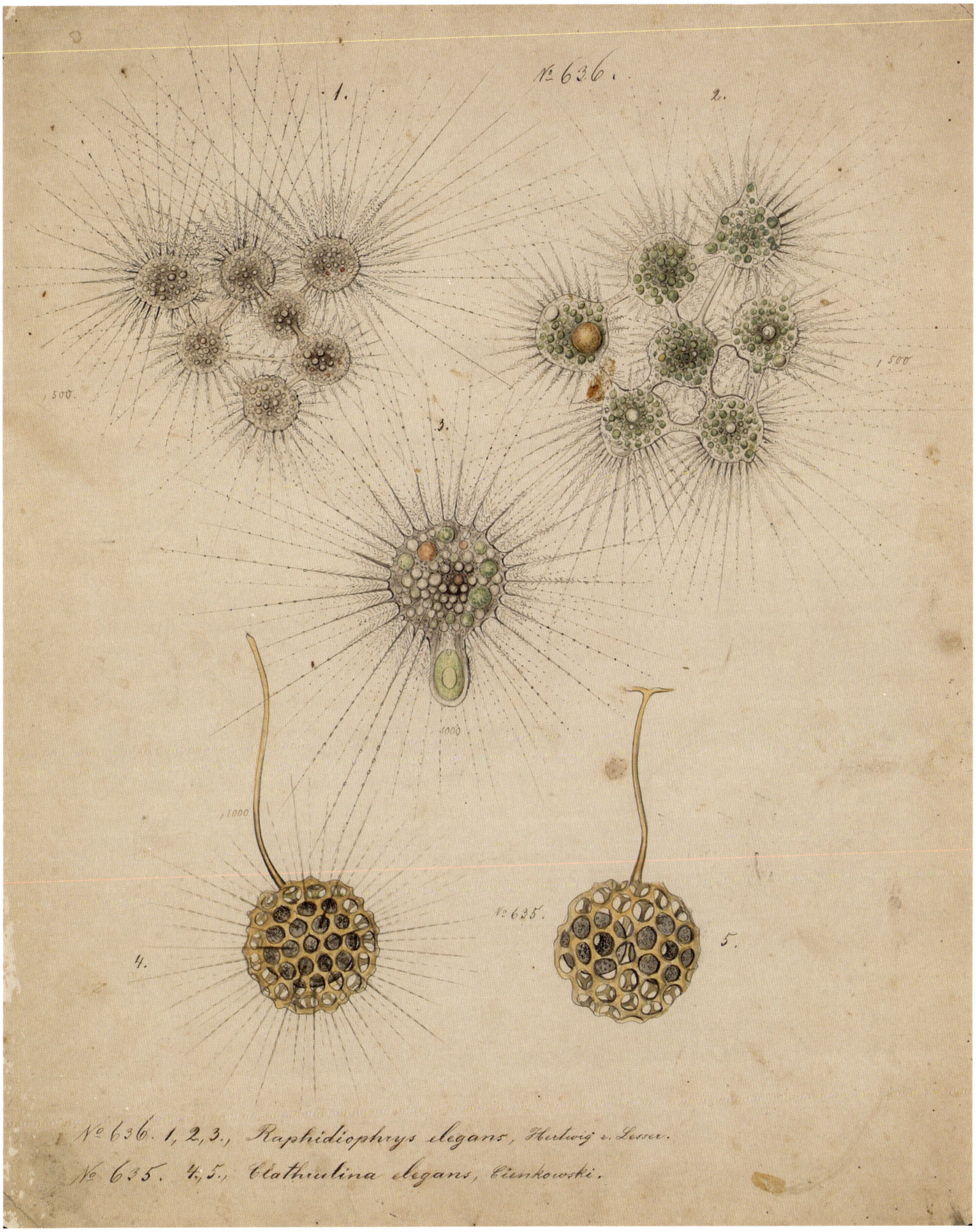
№ 636.
1.
2.
3.
4.
№ 635.
5.
№ 636. 1, 2, 3., Raphidiophrys elegans, Hertwig v. Lesser.
№ 635. 4, 5., Clathrulina elegans, Cienkowski.

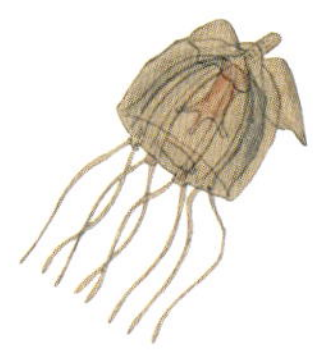

2.
Bougainvillia fruticosa Allman
Sea Fir (Hydroid)
Order: Athecata
Model 136
Pencil, ink, and watercolor on paper
42 cm x 32 cm

Hydroids belong to the Hydrozoa, a class of mostly small animals consisting of more than 2,700 species. Hydroids were so named in the 19th century because of their ability to grow new "heads" to replace heads that are lost or damaged. In Greek mythology, the Hydra was a monster with many heads that multiplied whenever one was cut off.

This drawing is a reversed copy of plate IX in George James Allman, *A Monograph of the Gymnoblastic or Tubularian Hydroids in Two Parts. Part I: The Hydroida in General*, London: The Ray Society, 1871.

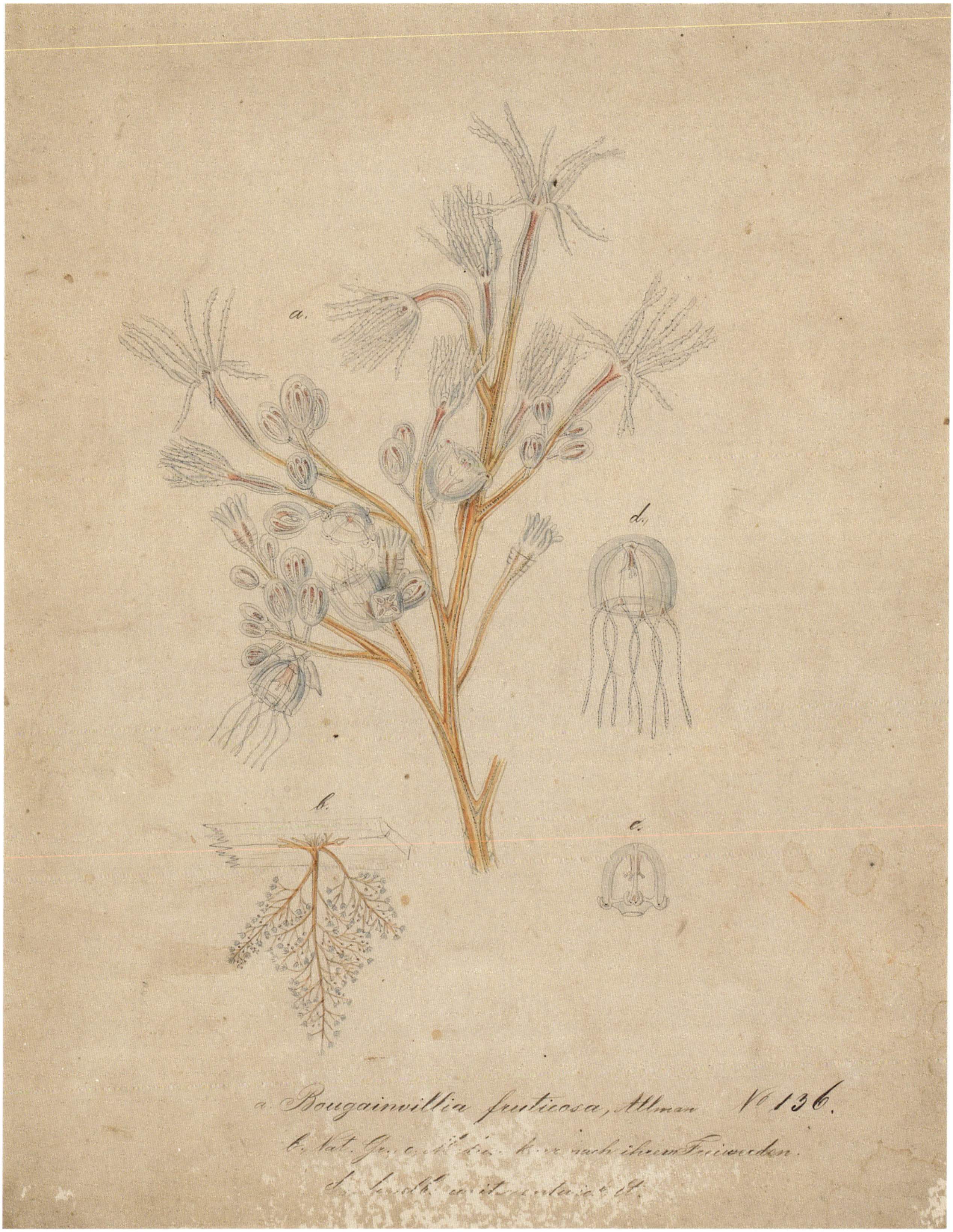
a.
d.
b.
c.
a. Bougainvillia fruticosa, Allman
No. 136.

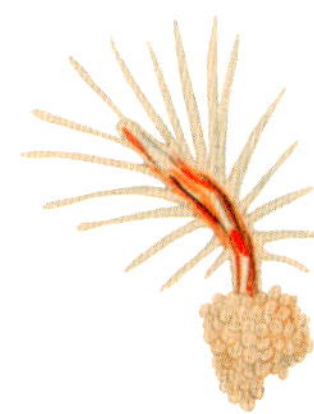

3.
Clava squamata Müller
Hydroid
Order: Athecata
Model 142

Pencil, ink, and watercolor on paper
40 cm x 32 cm

In its sessile (attached) stage, the body of a hydroid consists of a tube, closed at the bottom but with an opening at the top, through which the animal both takes in food and expels waste. A ring of tentacles surrounds the opening, and the hydroid uses these to capture its prey.

This drawing shows one view of the complete animal and six details.

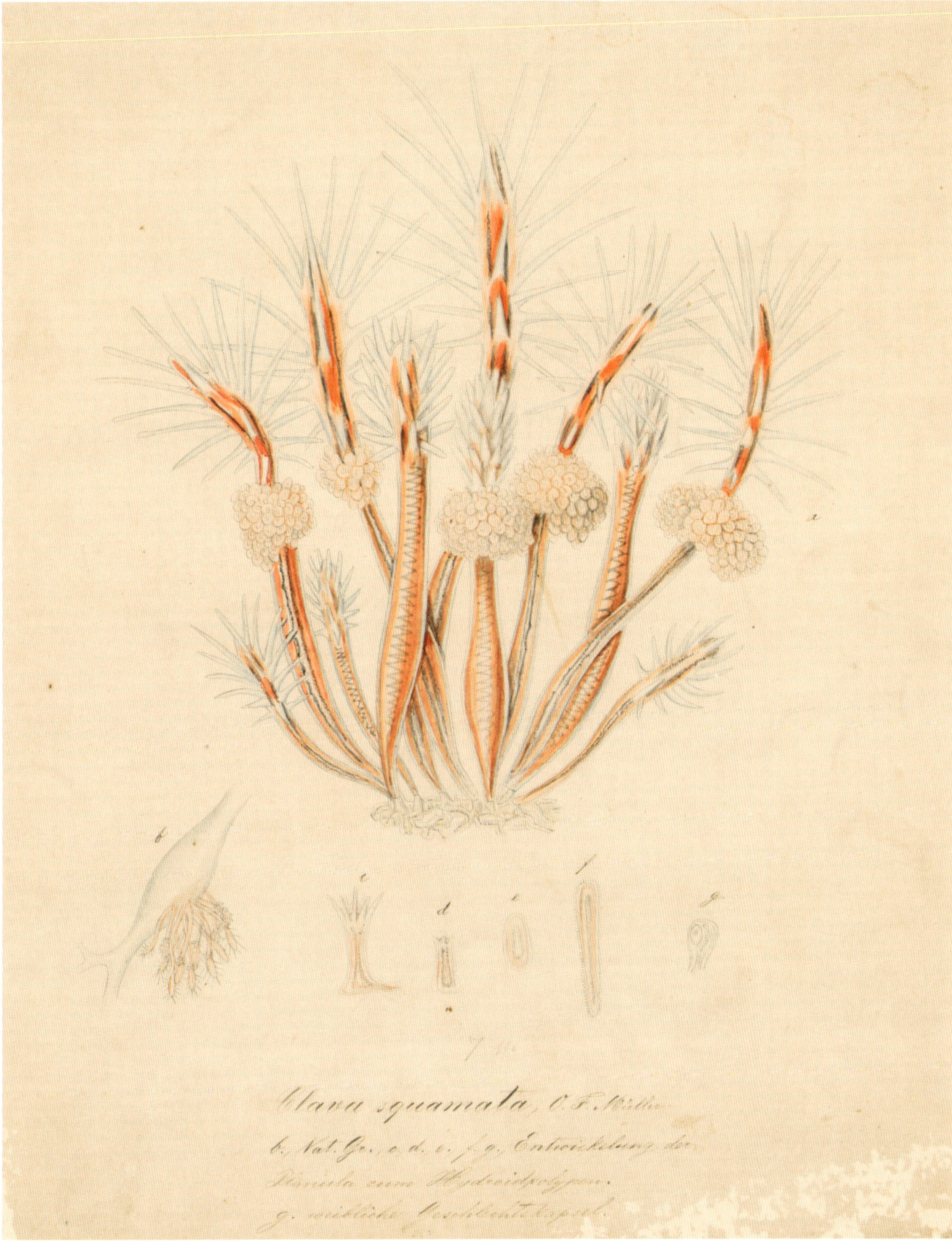
Clava squamata, O. F. Müller
b. Nat. Gr., c. d. e. f. g. Entwickelung der
Planula zum Hydroidpolypen.
g. weibliche Geschlechtskapsel.

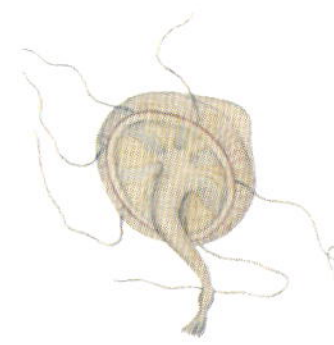

4.
Carmarina hastata Haeckel
Medusae
Order: Athecata
Model 138
Ink and watercolor on paper
41 cm x 33 cm

Hydroids have a life cycle of three stages. A tiny, free-swimming larva develops into a sessile (attached) animal, and this, in turn, releases a medusa (small jellyfish). Most hydroids live in marine environments, but some are found in fresh water.

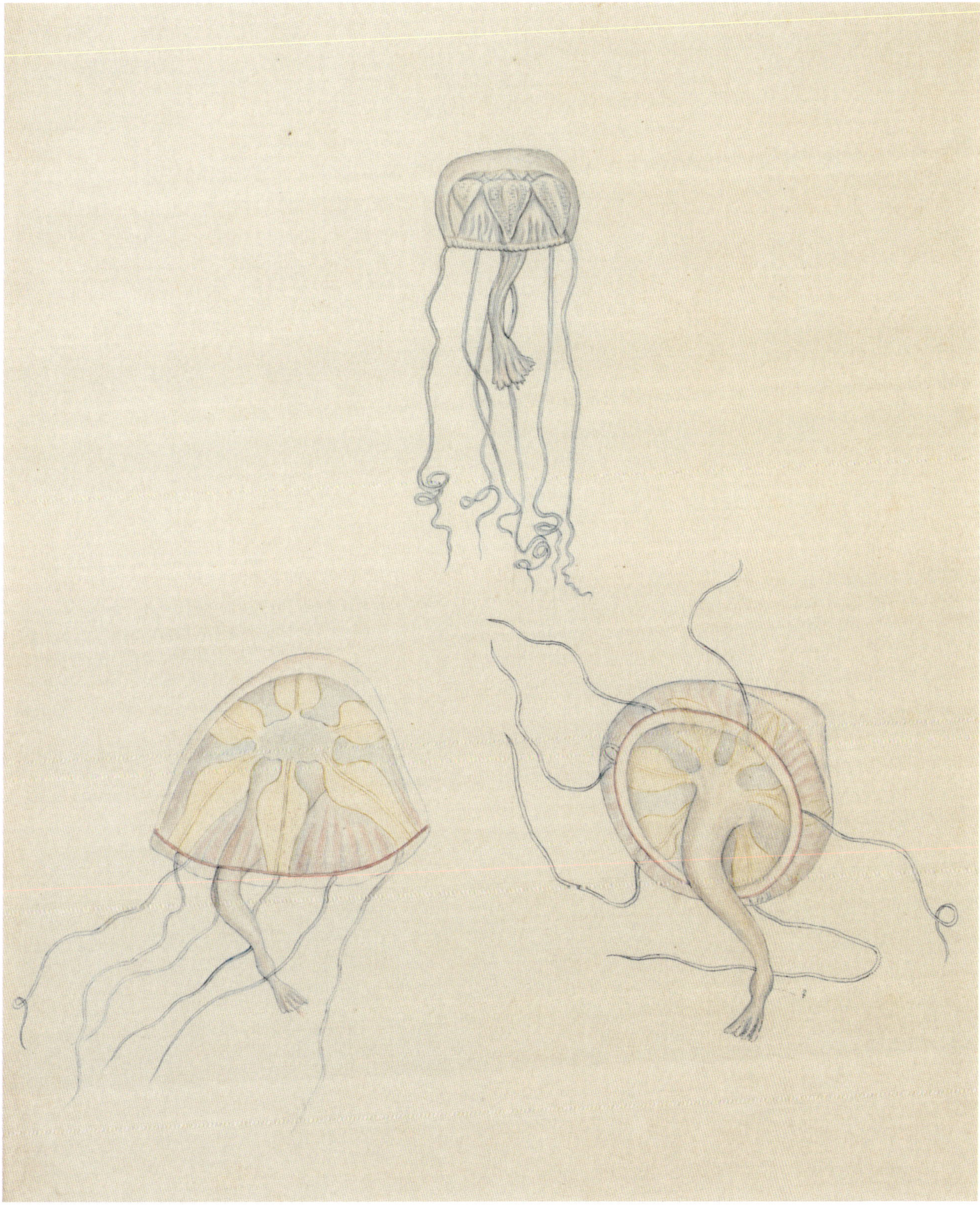

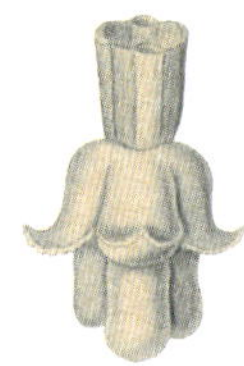

5.
Glossocodon eurybia Haeckel
Hydroid
Suborder: Trachomedusae
Model 153
Engraving on paper
28 cm x 34 cm

This engraving, made from a drawing by the great German zoologist Ernst Haeckel (1834–1919) and discovered among the Blaschkas' papers at Hosterwitz, was one of the many printed resources used by Leopold and Rudolf Blaschka to create their models of invertebrates. The image was plate II in Haeckel's *Beiträge zur Naturgeschichte der Hydromedusen. Erstes Heft: Die Familie der Rüsselquallen & 6 Kupfertafeln* (Contributions on the natural history of the Hydromedusae. First volume: The proboscidian jellyfish family [Medusae Geryonidae] with 6 copperplate engravings), Leipzig: Wilhelm Engelmann, 1865.

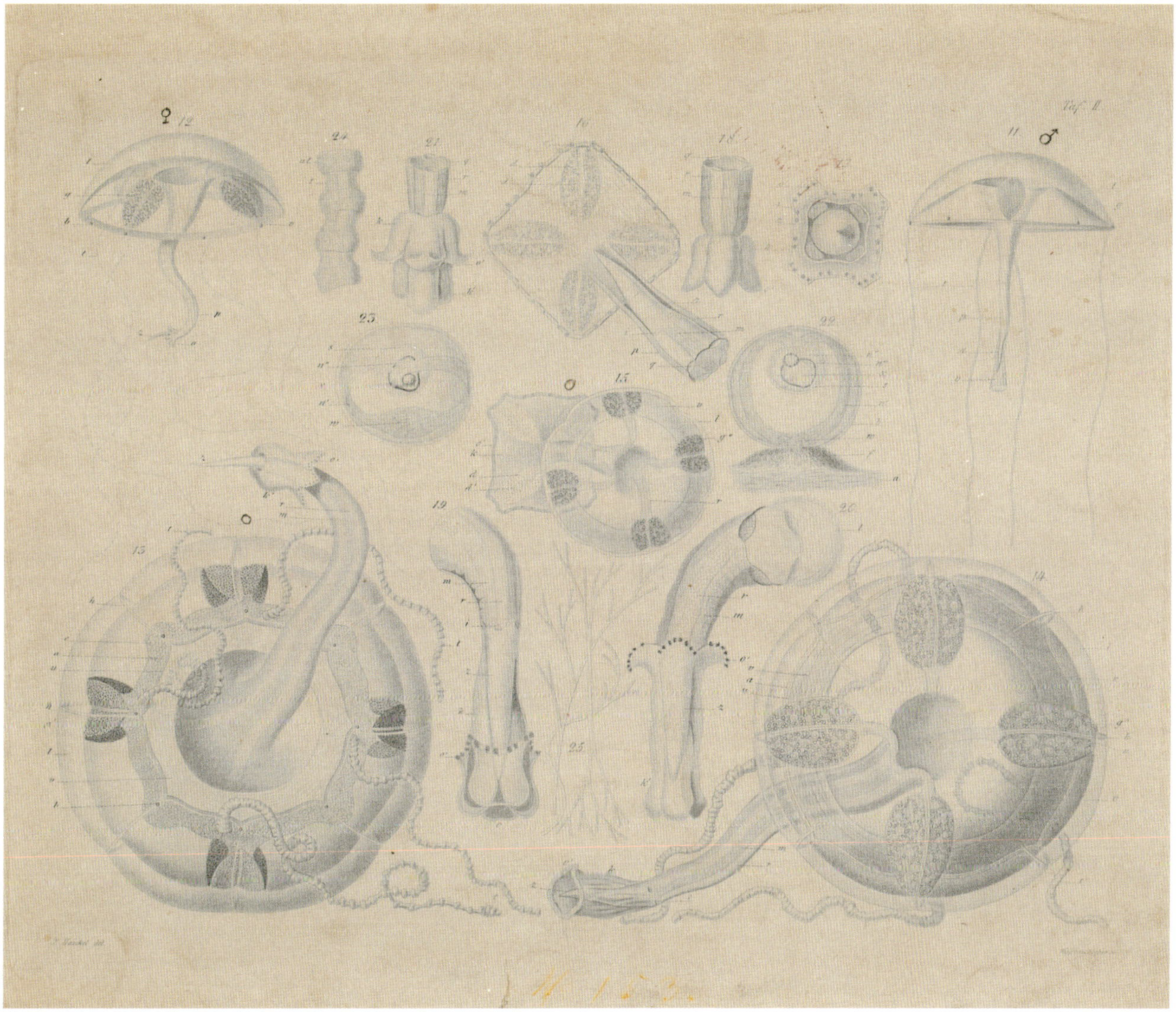

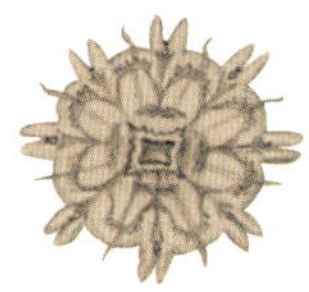

6.
Aurelia aurita Linnaeus
Common Jellyfish
Model 224
Cyanea capillata Linnaeus
Jellyfish
Model 231
Order: Semaeostomeae
Pencil on paper
24 cm x 29 cm

Jellyfish of the order Semaeostomeae are found in coastal waters worldwide, and the genera *Aurelia* and *Cyanea* are widely distributed around the coasts of North America and Europe. The familiar, free-swimming form of these animals is just one of the stages in their life cycle. This begins when the polyp, a small tubular organism, produces larvae. The larvae grow into adult jellyfish that are male or female, but they can change their gender. The adults mate, and the embryos are carried inside the females or develop in the sea.

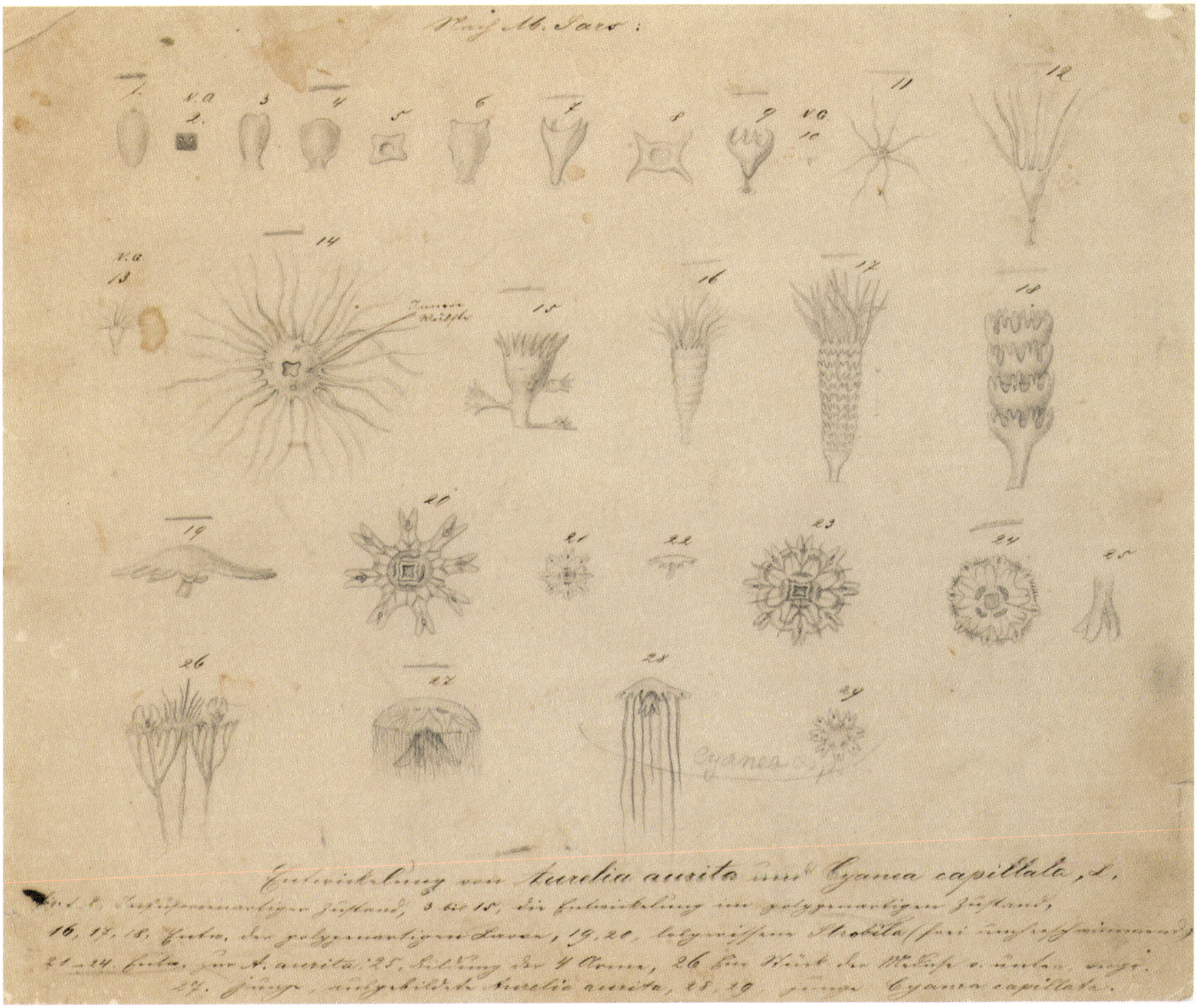
Cyanea
Entwickelung von Aurelia aurita und Cyanea capillata, L.

7.
Crambessa tagi Haeckel
Scyphozoan Jellyfish
Order: Rhizostomae
Model 230
Pencil, ink, and watercolor on paper
41 cm x 32 cm

About 200 species of scyphozoan jellyfish are known worldwide. One-quarter of these are found in Australian waters. Jellyfish of the order Rhizostomae do not have a central mouth, but instead they have numerous mouths on the eight armlike appendages that extend from the bottom of their umbrella-shaped body.

Crambessa tagi is one of some 4,000 species named by Ernst Haeckel, a scholar of enormous energy and an early champion of Charles Darwin's theory of evolution.

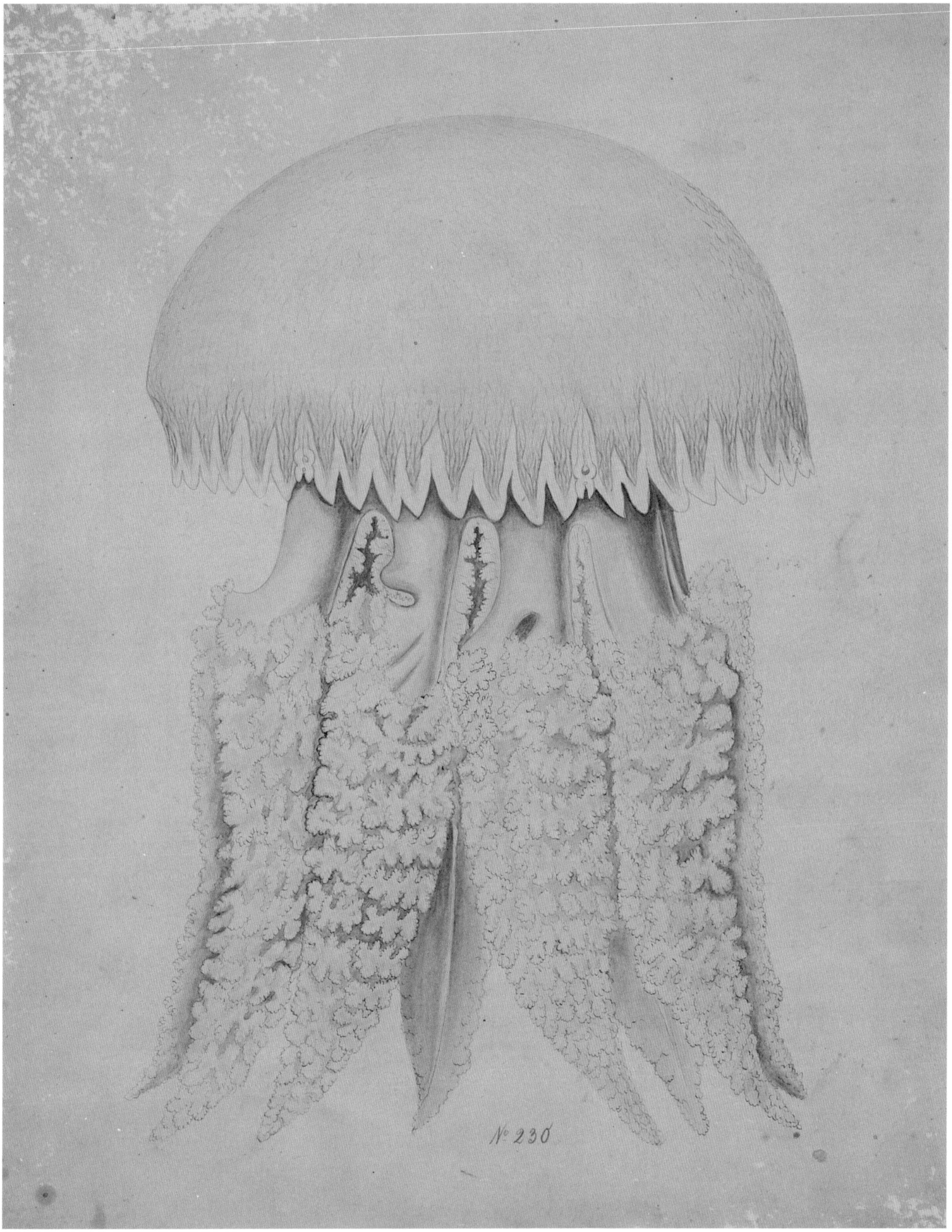
No 230

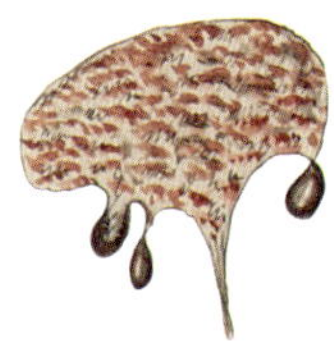

8.
Jellyfish (unidentified)
Pencil, ink, and watercolor on paper
32 cm x 41 cm

Jellyfish are almost 99 percent water. Most of them are carnivorous, capturing small animals in their tentacles, which have cells laden with toxins that disable their prey. Some, however, feed by filtering microscopic animals and plants from the water in which they swim. With few exceptions, jellyfish live for only a few weeks. They are found, sometimes in large numbers, drifting offshore, and they are frequently stranded on the beach by receding tides.

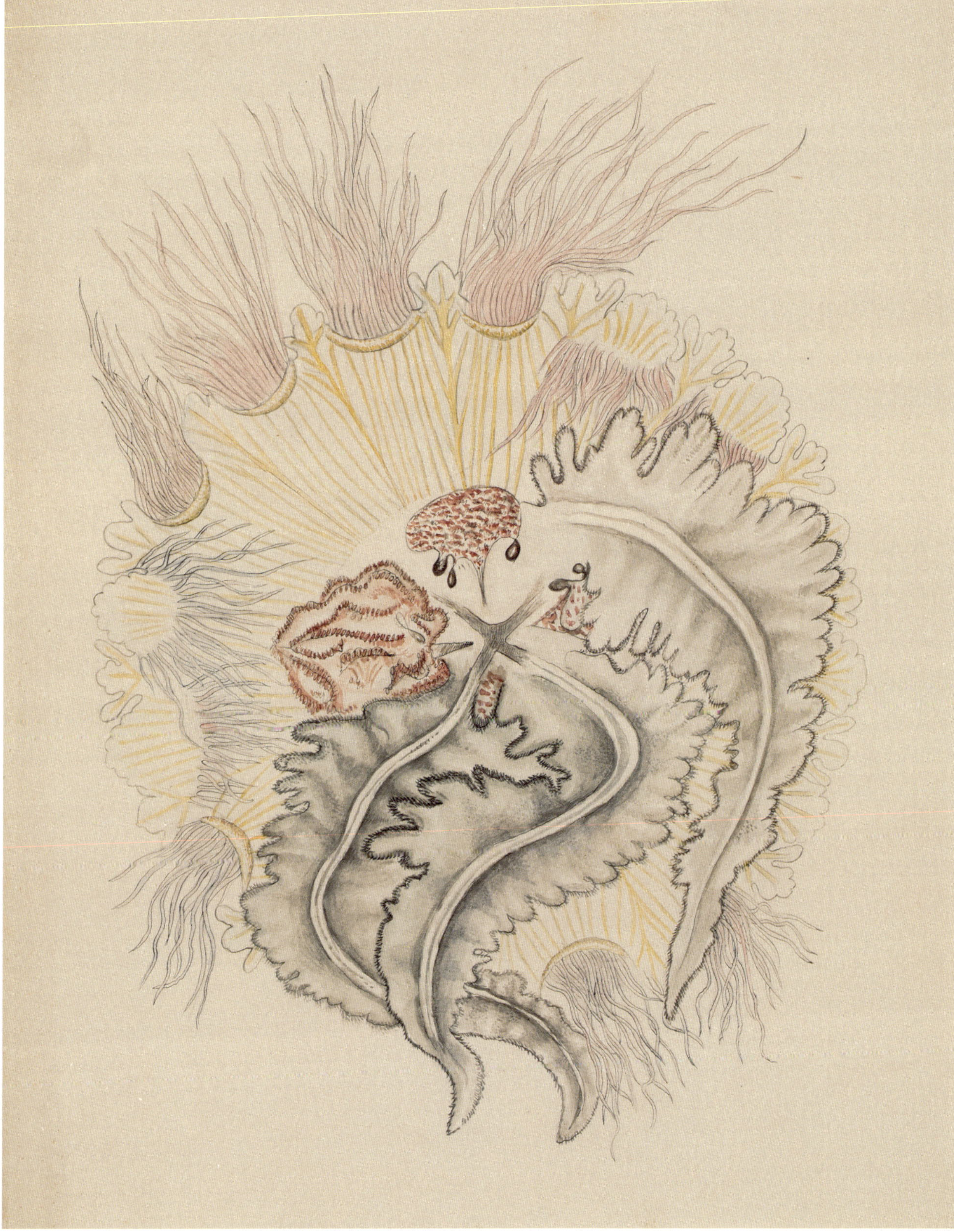

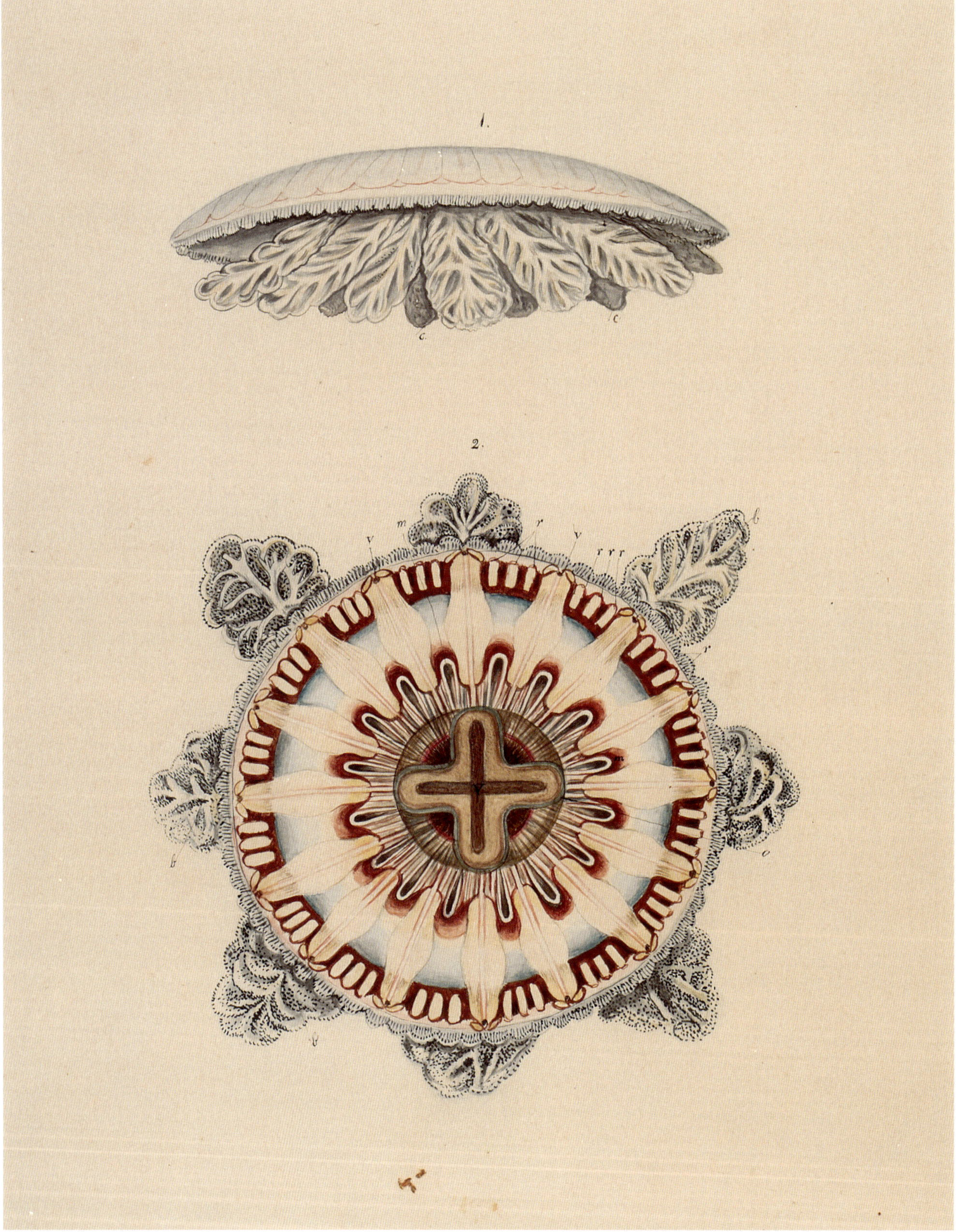

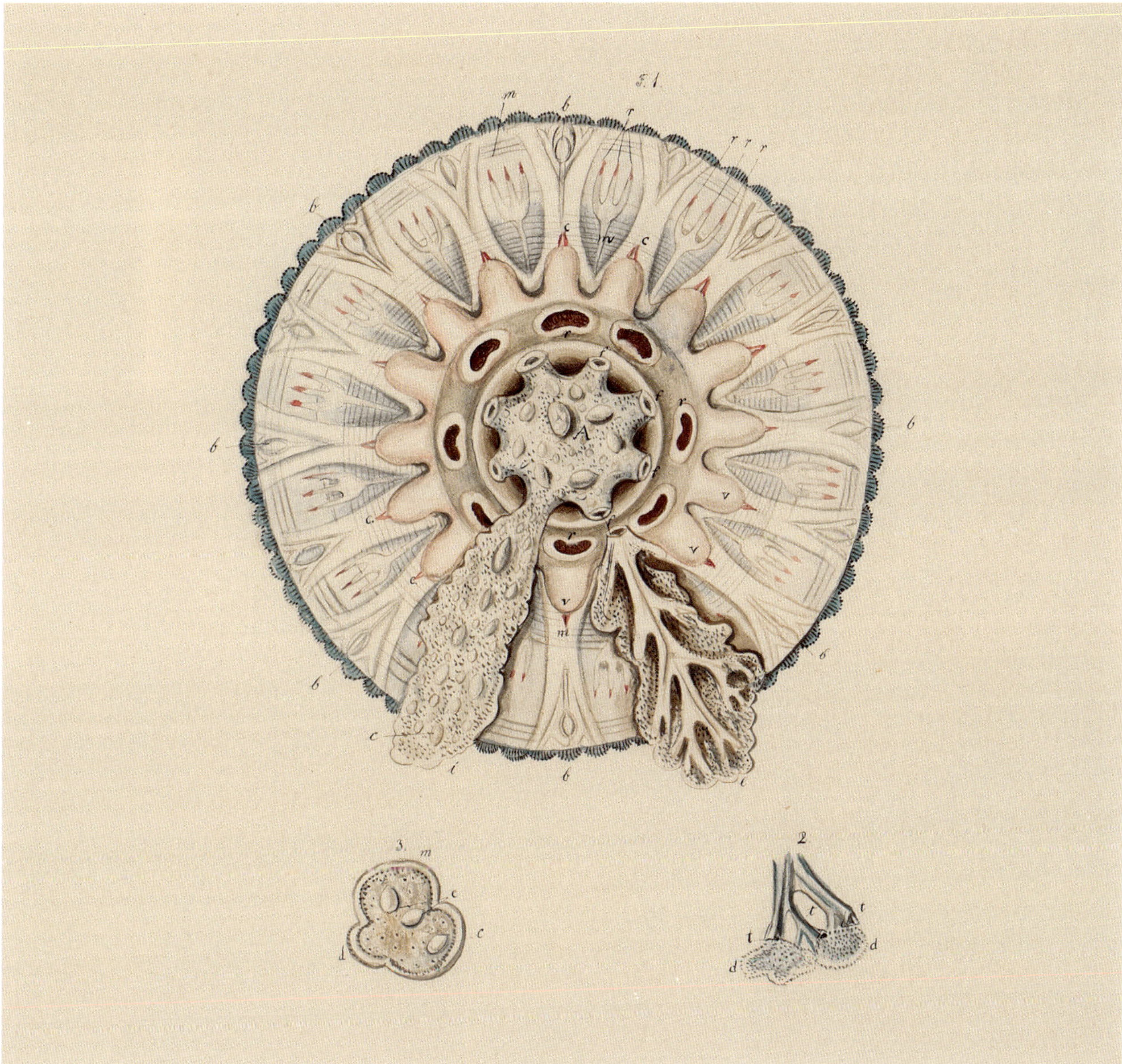

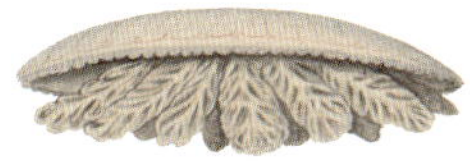

9.
Jellyfish (unidentified)
Pencil, ink, and watercolor on paper
41 cm x 64 cm

Although most jellyfish vary in size between about three and 40 centimeters, some species are much larger—up to almost two meters or more in diameter. Jellyfish are found in oceans worldwide.

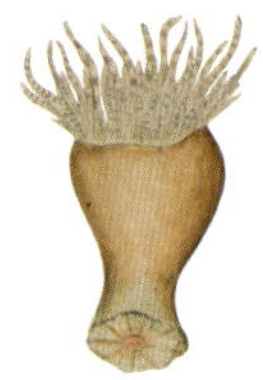

10.
Above and center:
Sagartia bella var. *punctata*
Schmarda
Sea Anemone
Model 89
Below:
Actinia chromatodera
Schmarda
Sea Anemone
Model 25
Order: Actiniaria
Pencil, ink, and watercolor on paper
39 cm x 32 cm

In his *A Naturalist's Rambles on the Devonshire Coast* (1853), Philip Henry Gosse described his difficulty in attempting to retrieve a specimen of *Sagartia bella*, which he called the daisy anemone. He wrote: "In the sunshine of a fair day, they expand beautifully. . . . But no sooner do the fingers touch one, than its beautifully circular disk begins to curl . . . receding . . . from the assault. Presently you discover that you can no longer touch it at all. . . . Nothing will do but the chisel."

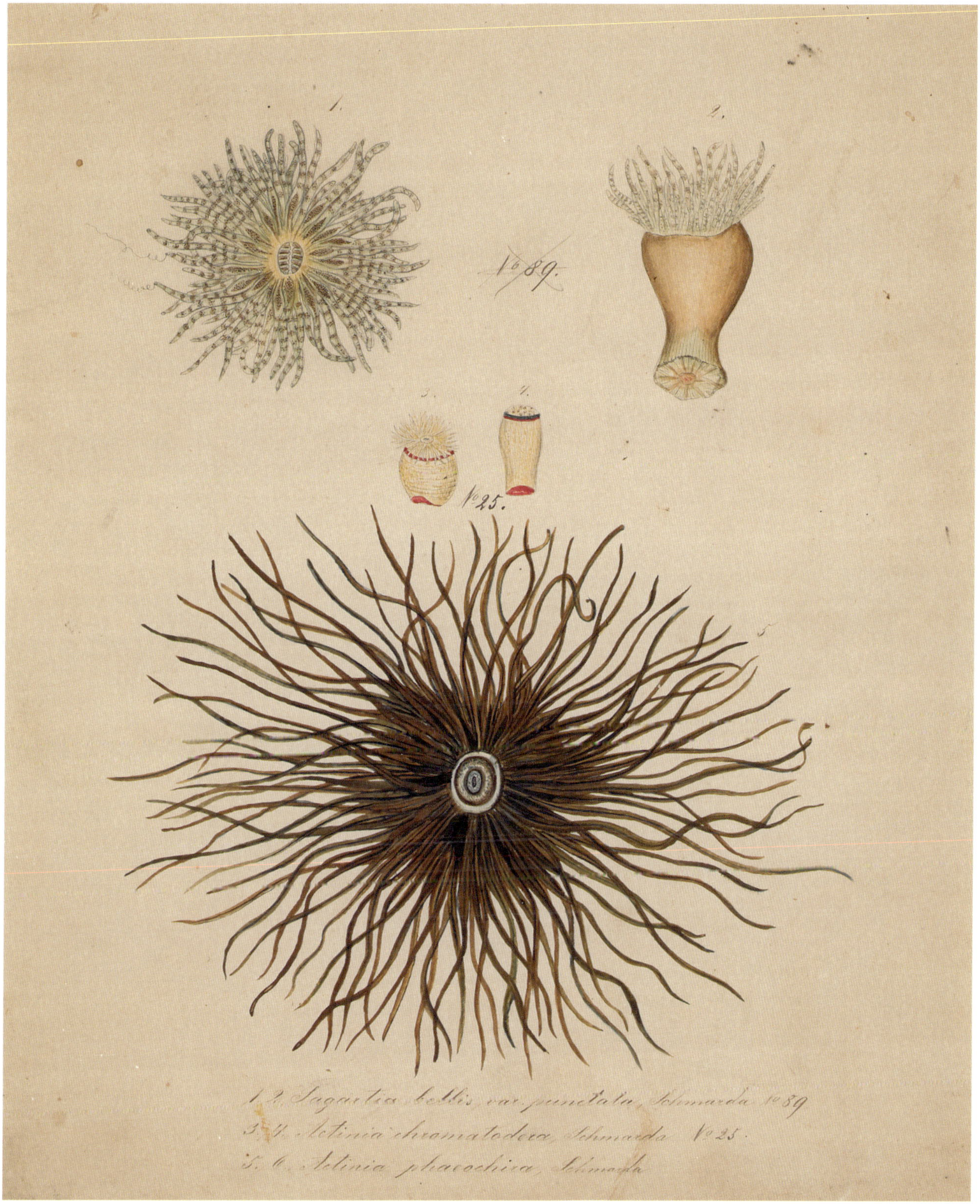

1.
2.
No 89.
No 25.
1.2. Sagartia bellis var. punctata, Schmarda No 89
3.4. Actinia chromatodera, Schmarda No 25.
5. 6. Actinia phaeochira, Schmarda

11.
Actinoloba dianthus Ellis,
var. *rubida* Gosse
Sea Anemone
Order: Actiniaria
Model 27
Watercolor on paper
39 cm x 33 cm

Sea anemones, which were among the first marine invertebrates created in glass by Leopold Blaschka, are so called because of the flowerlike appearance of the tentacles that surround the mouth. They normally frequent rocky shores, where most species attach themselves to rocks or sturdy weeds. These animals have a worldwide distribution, but they are more common in warm than in colder waters.

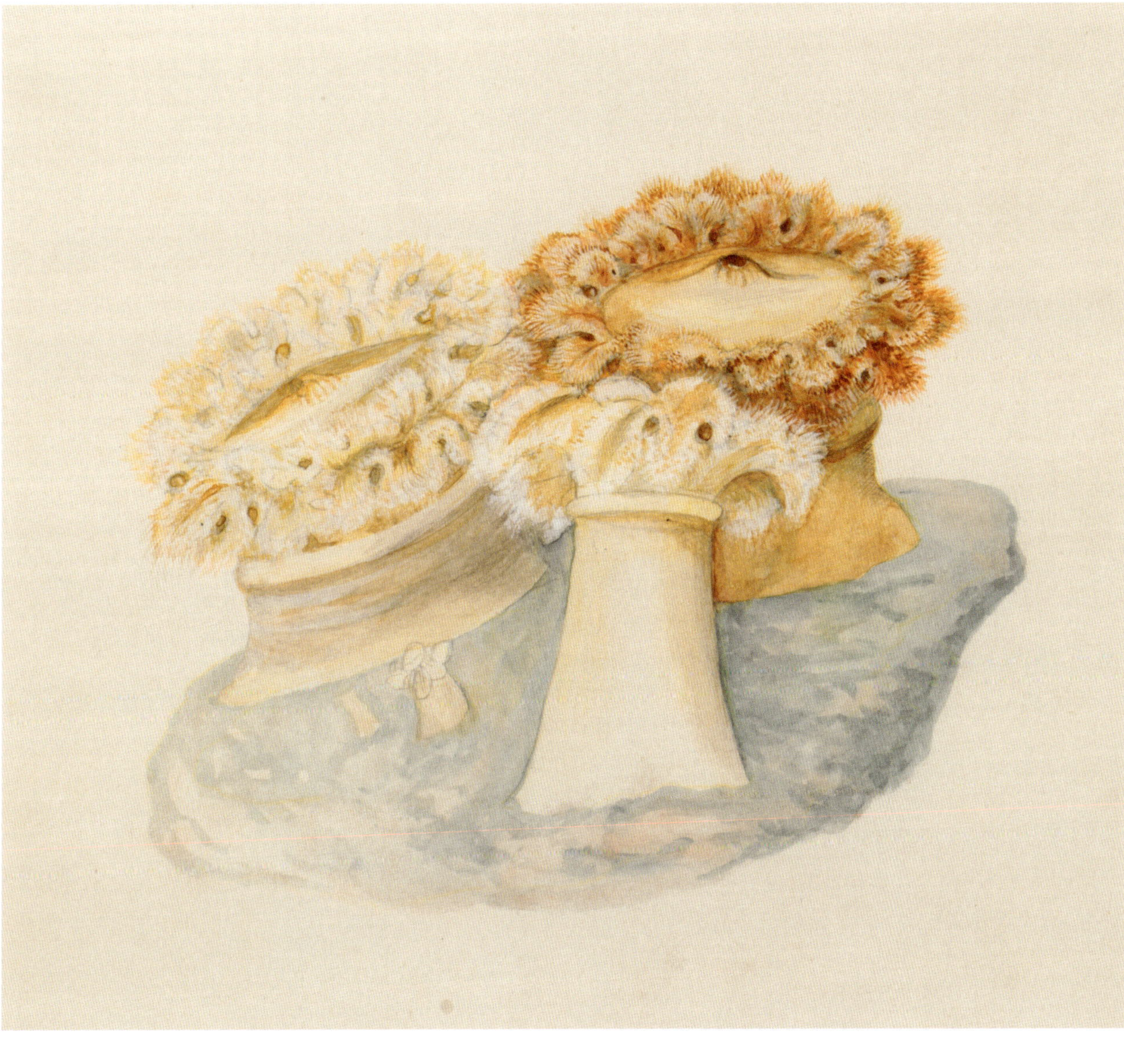

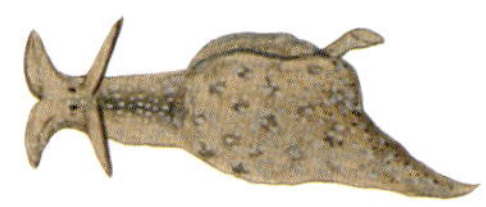

12.
Syphonota punctata Pease
Model 500
Syphonota viridescens Pease
Model 501
Pleurobranchus grandis Pease
Model 496
Comb Jellies
Order: Cydippida
Pencil, ink, and watercolor on paper
41 cm x 32 cm

The comb jelly is the largest animal that uses cilia for locomotion. Its eight rows of combs consist of plates with thousands of cilia that beat in order to move the animal forward and to draw food and water into its oral canals. When comb jellies luminesce at night, their combs can be seen beating.

These drawings were copied by Leopold Blaschka from an illustration in W. H. Pease, "Descriptions of Sixty-Five New Species of Marine Gasteropodae Inhabiting Polynesia," *American Journal of Conchology*, v. 3, 1868, pp. 271–297.

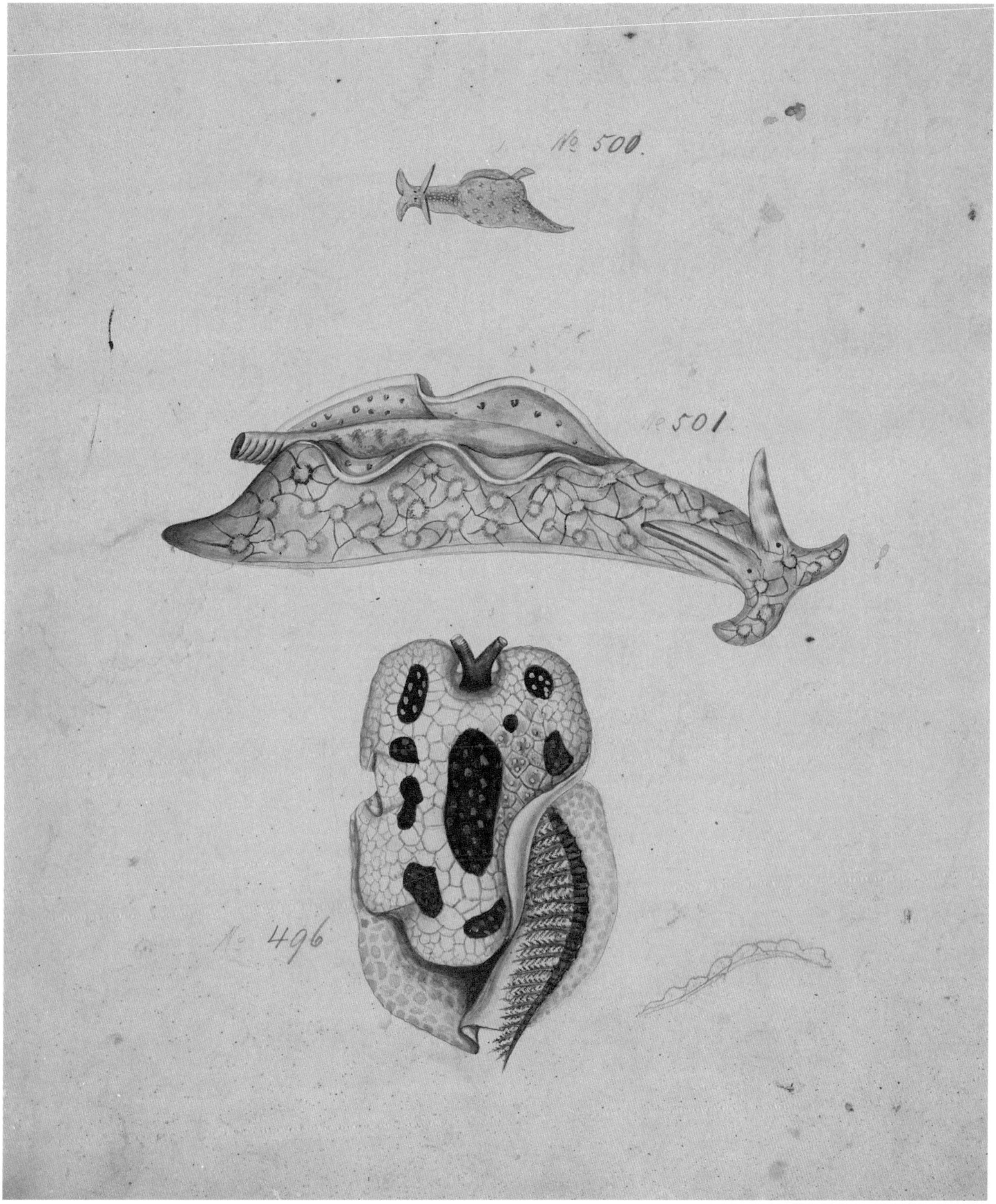
№ 500.
№ 501.
№ 496

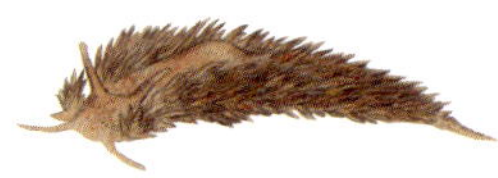

13.
Above:
Aeolis papillosa Linnaeus
Model 369
Below:
Aeolis exigua Alder and Hancock
Model 365
Sea Slugs
Subclass: Opisthobranchia
Pencil, ink, and watercolor on paper
41 cm x 32 cm

Although they are among the mollusks, a vast phylum of about 130,000 species, most of which have shells, sea slugs do not have shells. These creatures are often brightly colored, and the dorsal (upper) surface of their body is ornamented with fleshy protrusions known as cerata. The head has two antennalike organs.

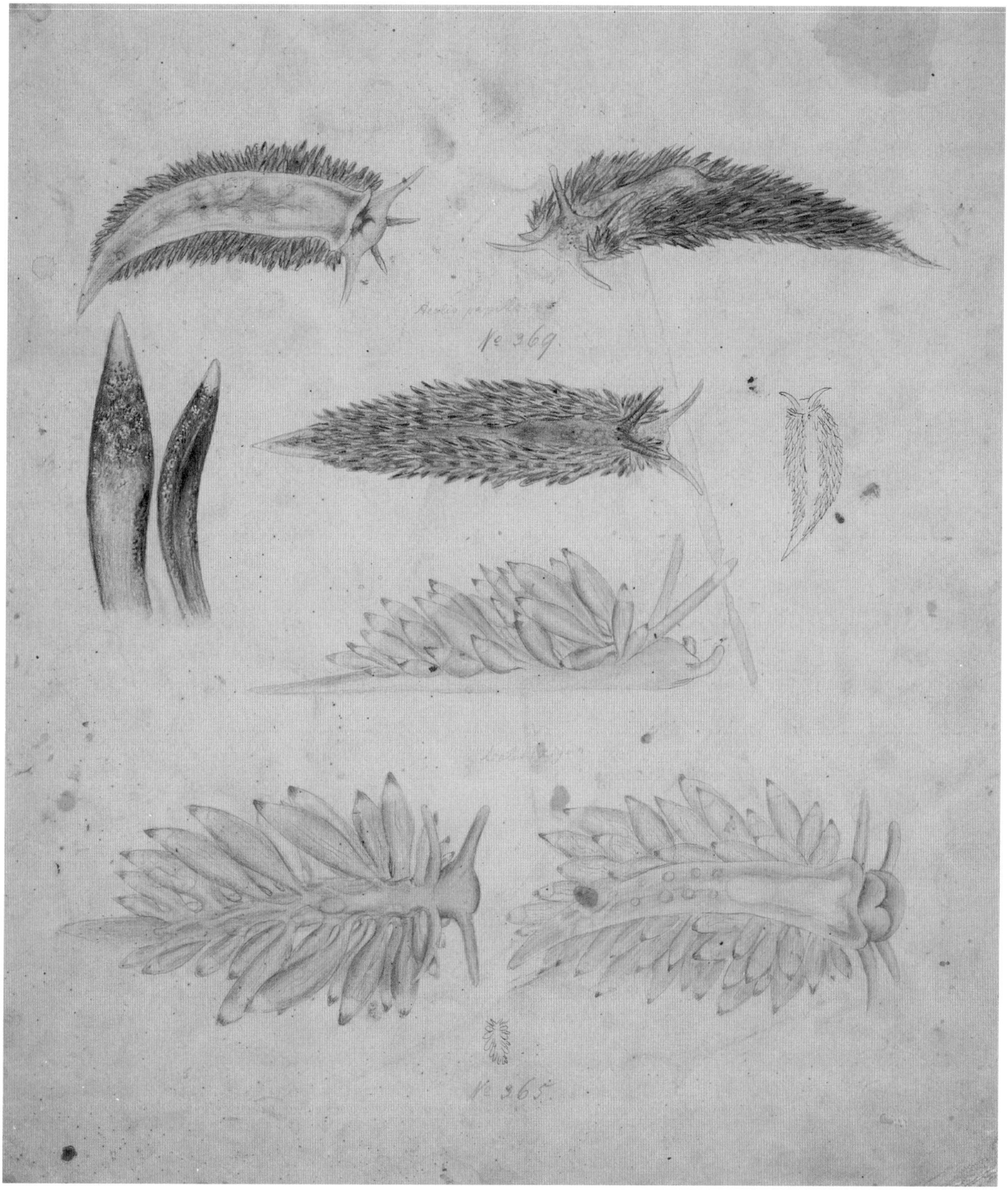
No 369
No 265

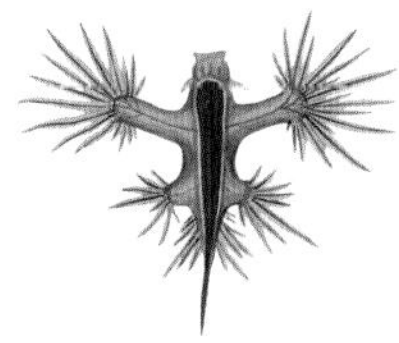

14.
Glaucus longicirrus Reinhardt
Model 450
Glaucus lineatus Reinhardt
Model 449
Glaucilla briareus Reinhardt
Model 446
Glaucilla marginata Reinhardt
Model 447
Sea Slugs
Order: Acoela
(= Nudibranchia)
Pencil, ink, and watercolor
on paper
41 cm x 32 cm

Sea slugs are found in the shallow waters of all of the world's oceans. They are carnivores, feeding on other invertebrate animals, such as sea anemones. Sea slugs of the genus *Glaucus* have a very wide distribution in warm environments.

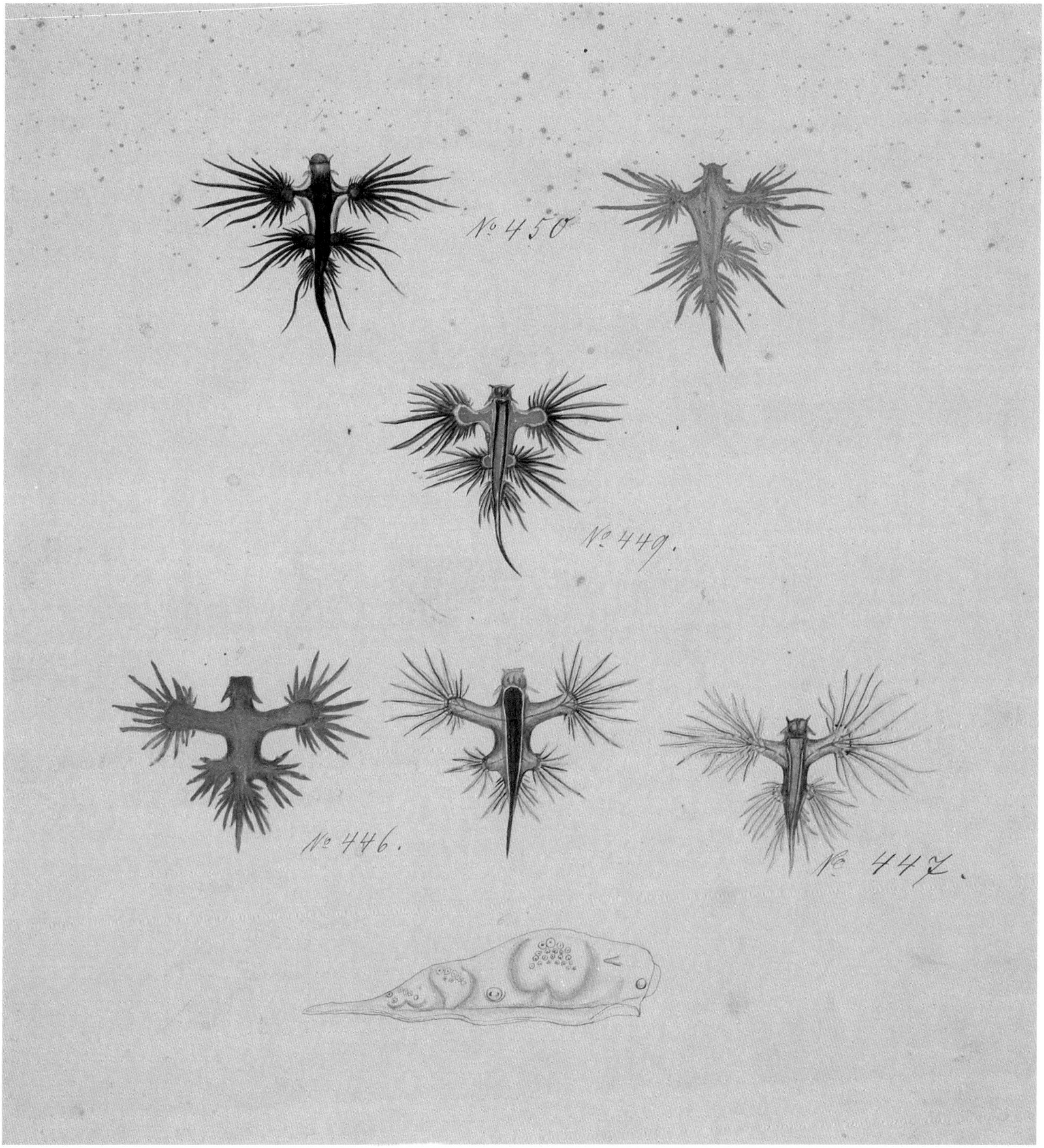

No 450
No 449.
No 446.
No 447.

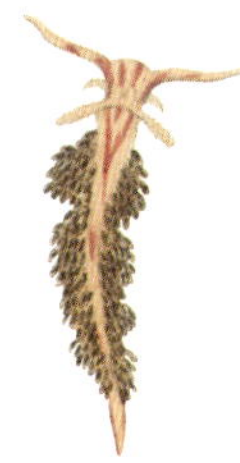

15.
Left to right:
Top row:
Flabellina newcombi
Model 444
Melibaea australis
Aeolis foulisi
Model 336
Elysia coodgeensis
Middle row:
Bornella hermanni
Flabellina ianthina
Model 443
Flabellina ornata
Model 445
Bottom row:
Janus sanguineus
Aeolis macleayi
Sea Slugs (all of the species were named by Angas)
Order: Acoela
(= Nudibranchia)
Engraving on paper
21.5 cm x 13 cm

This is plate VI in the publication by G. F. Angas, in which he described, illustrated, and named sea slugs from the vicinity of Port Jackson, New South Wales, Australia. The publication, "Description d'espèces nouvelles appartenant à plusieurs genres de mollusques nudibranches de Port-Jackson (Nouvelles-Galles du Sud), accompagnées de dessins faits d'après nature" (Description of new species of various kinds of nudibranchiate mollusks from Port Jackson [New South Wales], together with designs from nature), appeared in the *Journal de Conchyliologie*, v. 12, 1864, pp. 43–70. The plate is among the Blaschkas' papers in the Rakow Research Library of The Corning Museum of Glass. The Blaschkas produced models of four of the nine species shown on this plate.

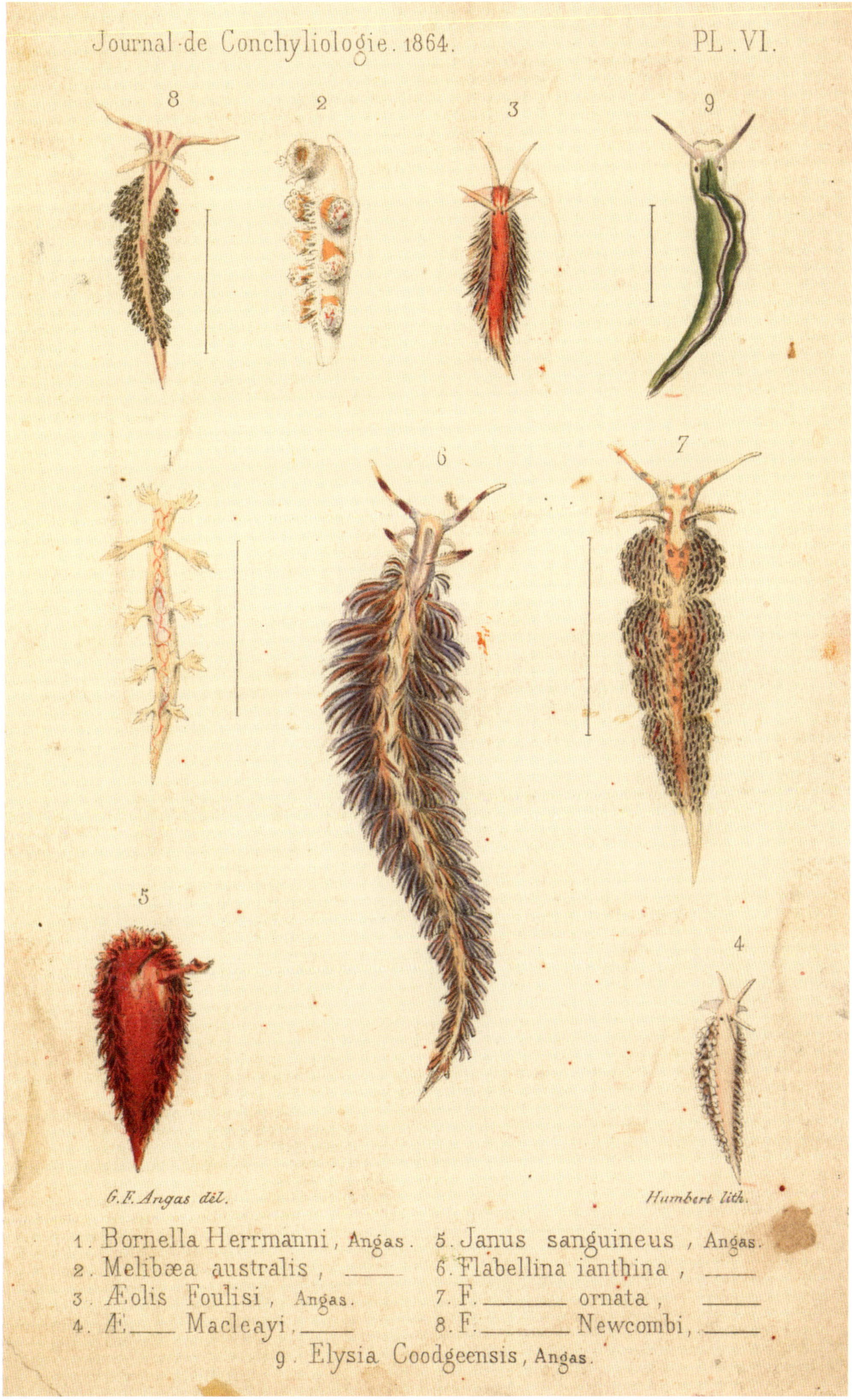
Journal de Conchyliologie. 1864.
PL. VI.
8
2
3
9
1
6
7
5
4
G. F. Angas del.
Humbert lith.
1. Bornella Herrmanni, Angas.
2. Melibæa australis, ——
3. Æolis Foulisi, Angas.
4. Æ. —— Macleayi, ——
5. Janus sanguineus, Angas.
6. Flabellina ianthina, ——
7. F. —— ornata, ——
8. F. —— Newcombi, ——
9. Elysia Coodgeensis, Angas.

16.
Hexabranchus praetextus
Ehrenberg
Sea Slug
Order: Acoela
(= Nudibranchia)
Pencil, ink, and watercolor
on paper
41 cm x 32 cm

The caption attached to this drawing contains an error (*praetexta* for *praetextus*), and it notes, correctly, that the species inhabits the Red Sea.

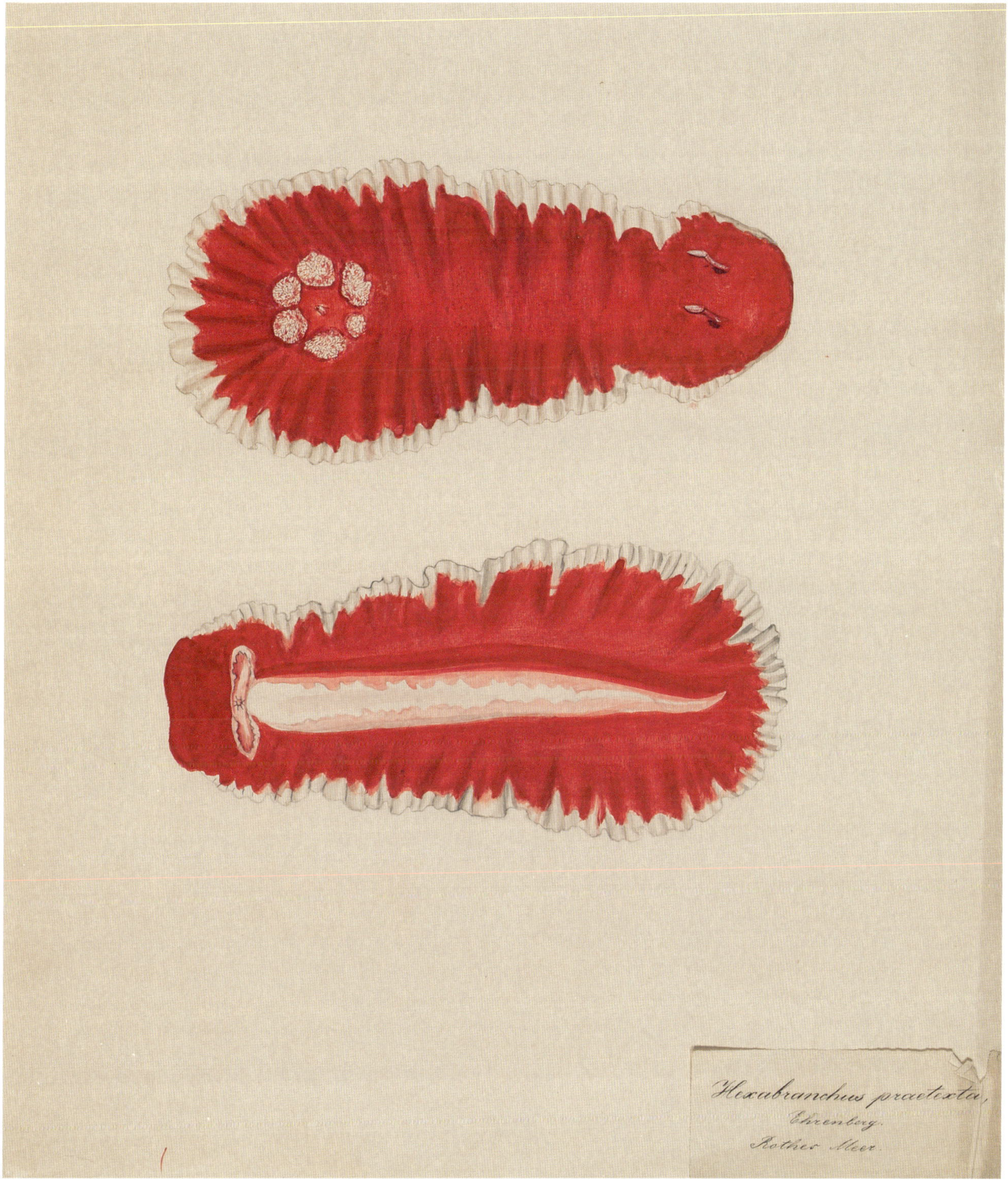
Hexabranchus praetexta,
Ehrenberg.
Rother Meer.

17.
Tethys leporina Linnaeus
Sea Slug
Order: Acoela
(= Nudibranchia)
Model 482
Ink and watercolor
on paper
25 cm x 32 cm

Most sea slugs, as their name implies, live on the sea floor. However, those of the family Tethyidae, which includes *Tethys leporina*, swim as well as creep along the bottom.

This drawing, on grayish brown paper, is unusual. Almost all of the other drawings in the Blaschkas' archive are on white or almost white paper.

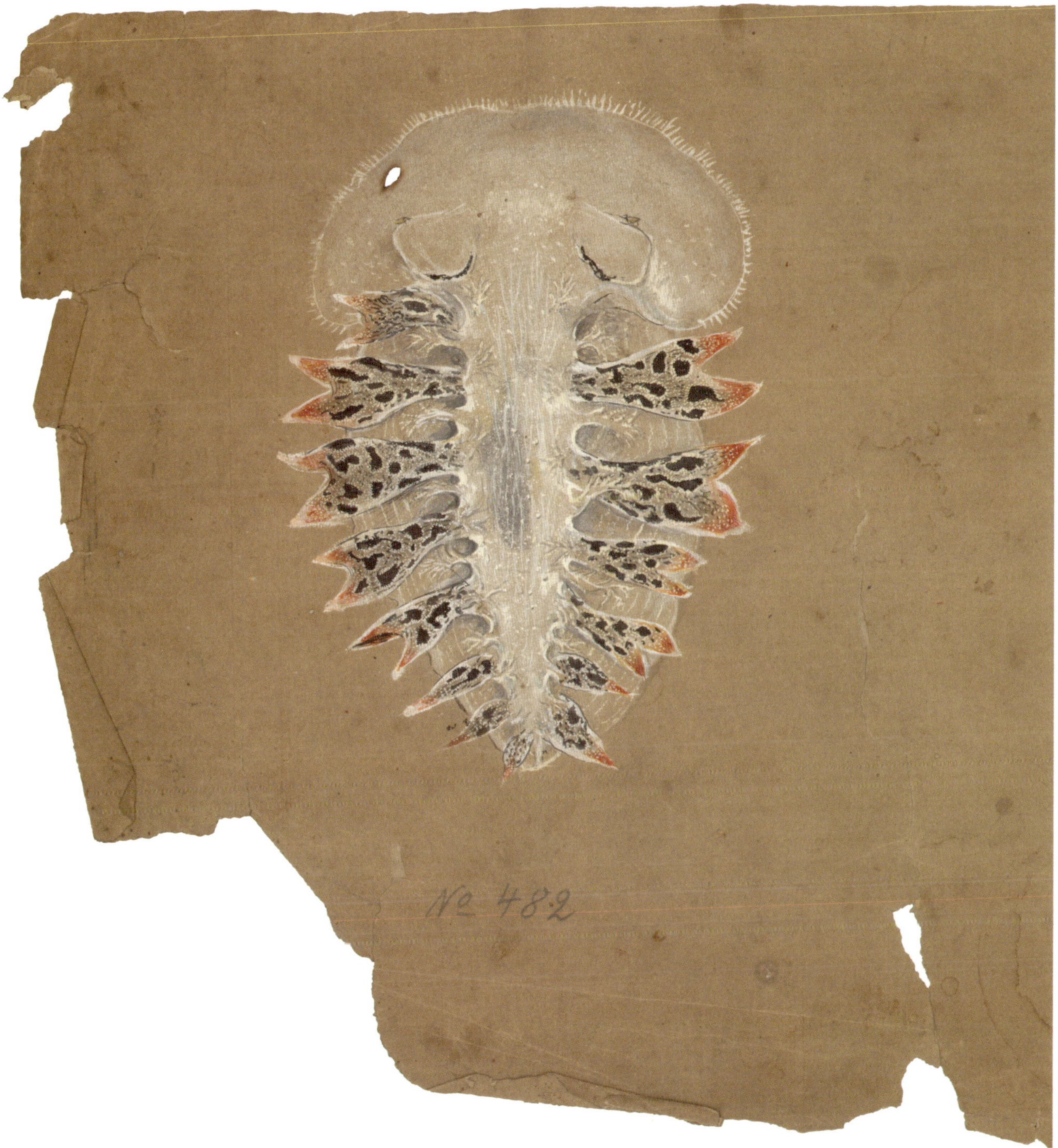
№ 482

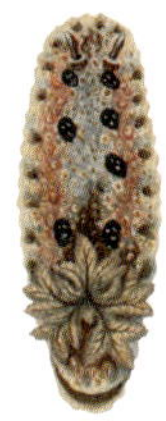

18.
Sea Slugs from India:
Doriopsis atromaculata
Alder and Hancock
(no. 20 in the illustration)
Model 397
Doriopsis clavulata
Alder and Hancock
(nos. 10 and 12)
Model 398
Doriopsis gemmacea
Alder and Hancock
(nos. 4 and 5)
Model 399
Doriopsis nigra Stimpson
(no. 13)
Model 400
Doriopsis rubra Kelaart
(no. 2)
Model 401
Order: Acoela
(= Nudibranchia)
Engraving on paper
25 cm x 32.5 cm

This is plate 31 from a groundbreaking paper on sea slugs published in London in 1866: Joshua Alder and Albany Hancock, "Notice of a Collection of Nudibranchiate Mollusca Made in India by Walter Elliot, Esq., with Descriptions of Several New Genera and Species," *Transactions of the Zoological Society of London*, v. 5. It provides a clear example of the scope of the scientific literature that Leopold Blaschka studied in order to expand the range of his models of marine invertebrates. The plate was found among the Blaschkas' papers in their house at Hosterwitz.

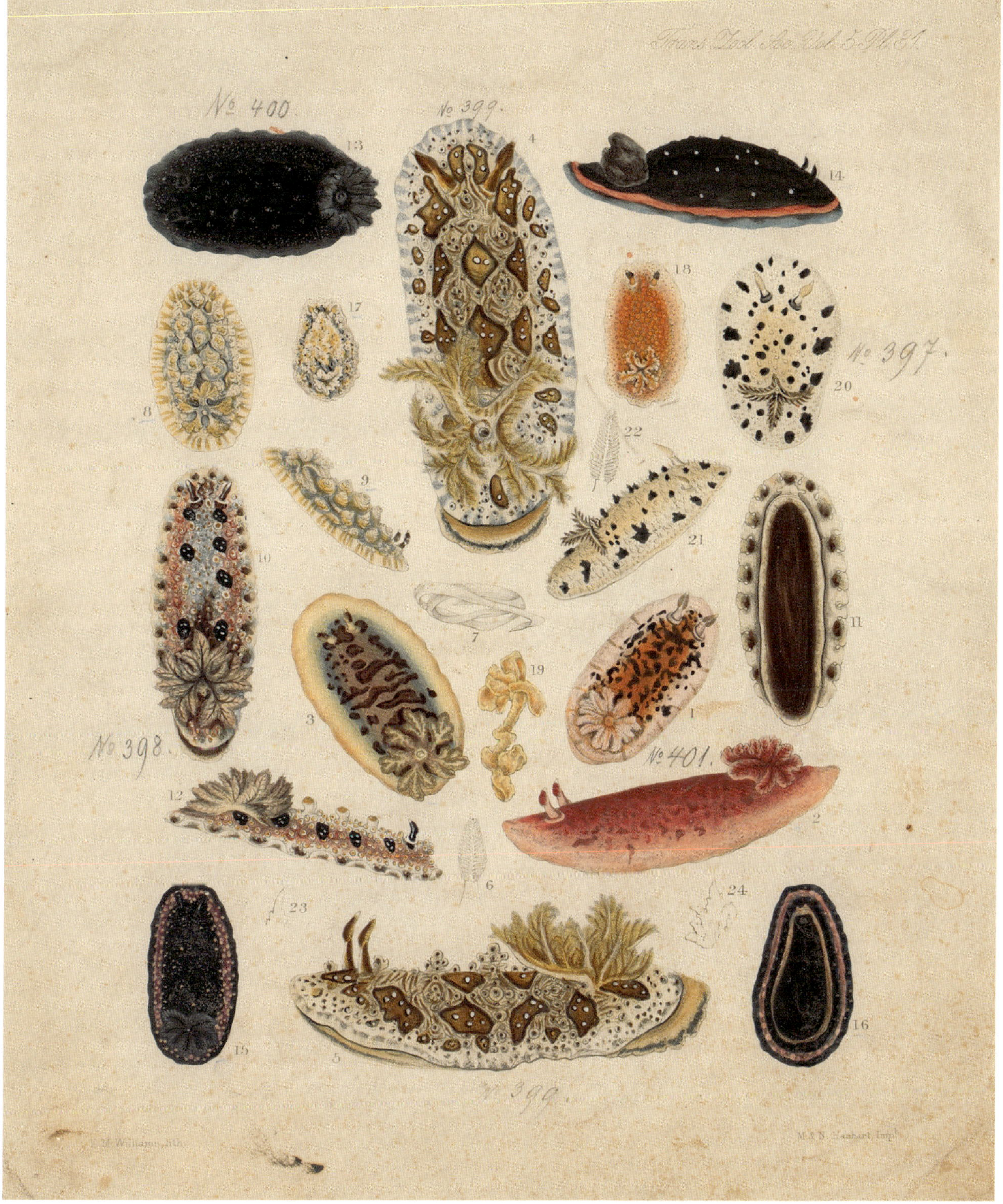

Trans. Zool. Soc. Vol. 5. Pl. 31.
No 400.
No 399.
No 397.
No 398.
No 401.
No 399.
E. M. Williams, lith.
M. & N. Hanhart, Imp.

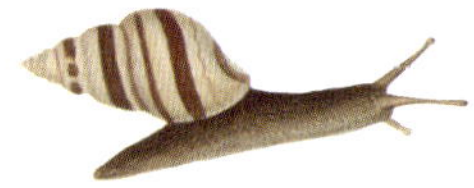

19.
Top to bottom:
Limax columbianus Gould
Slug
Arion foliolatus Gould
Slug
Glandina truncata Say
Snail
Bulimus fasciatus Müller
Snail
Order: Geophila
(= Stylommatophora)
Ink and watercolor on paper
41 cm x 33 cm

Unlike the overwhelming majority of the Blaschkas' drawings and models of invertebrate animals, which represent marine species, this careful drawing shows four different terrestrial creatures: two slugs and two snails. They belong to the Mollusca, the vast phylum of animals (comprising more than 128,000 species) that are found in the sea, in fresh water, and on land. Mollusks of the order Geophila have adapted to life on land by protecting themselves from desiccation and the cold by, for example, hibernating.

As far as we are aware, the Blaschkas sold models of only one of these animals: *Glandina truncata*, which was no. 699 in their latest trade catalog.

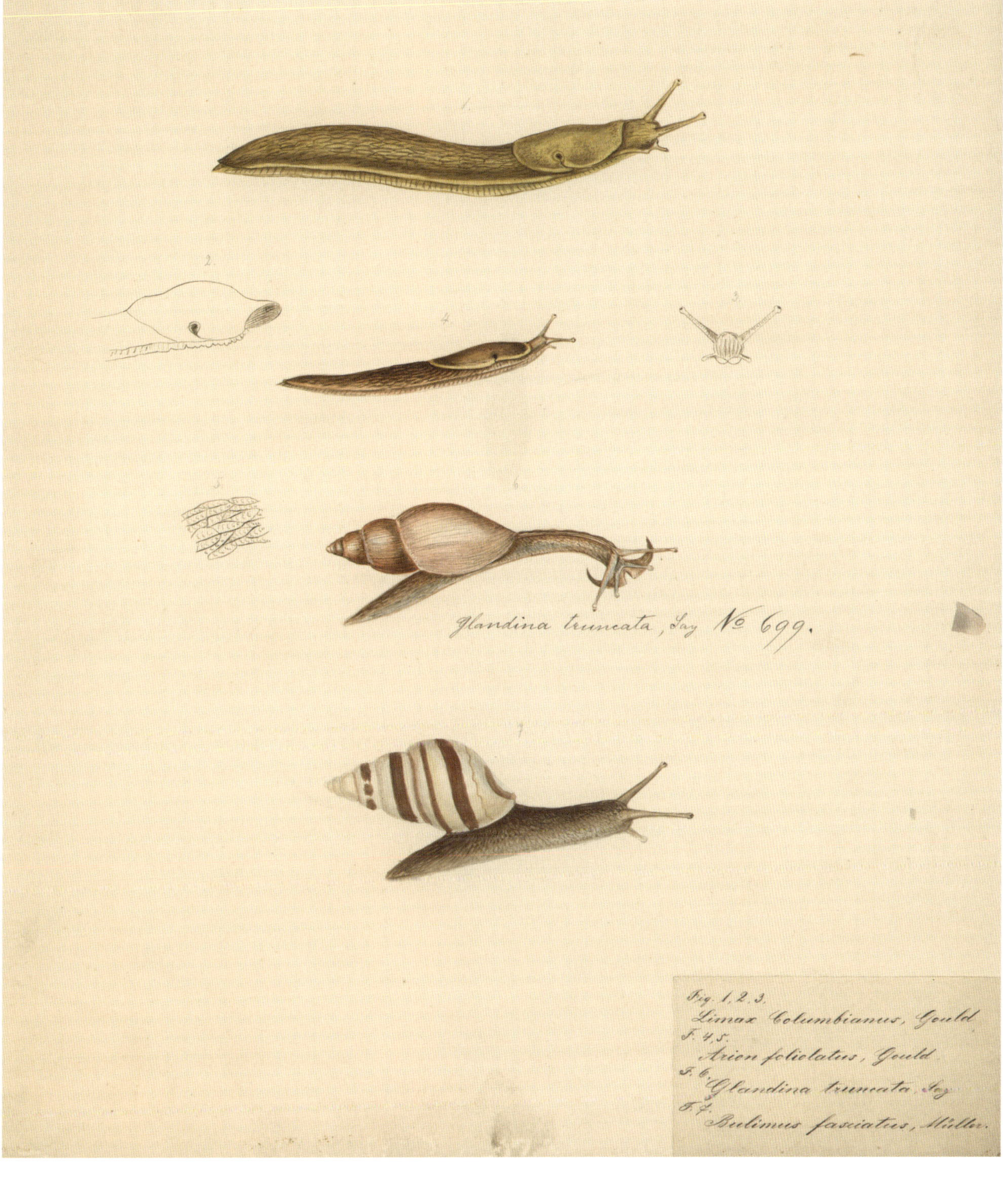

Glandina truncata, Say No 699.
Fig. 1, 2, 3.
Limax Columbianus, Gould
F. 4, 5.
Arion foliolatus, Gould
F. 6.
Glandina truncata, Say
F. 7.
Bulimus fasciatus, Müller.

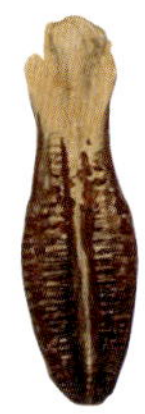

20.
Mytilus edulis Linnaeus
Blue Mussel
Class: Lamellibranchia
Model 673
Pencil, ink, and watercolor on paper
19 cm x 30 cm

Mussels inhabit both seawater and fresh water, and they are found all over the world. They attach themselves to solid objects, such as rocks and man-made structures, by threads, and they frequently occur in large, densely concentrated clusters. The blue mussels, which are either blue or deep purple, are a popular variety of seafood in Europe, where they have been cultivated for centuries.

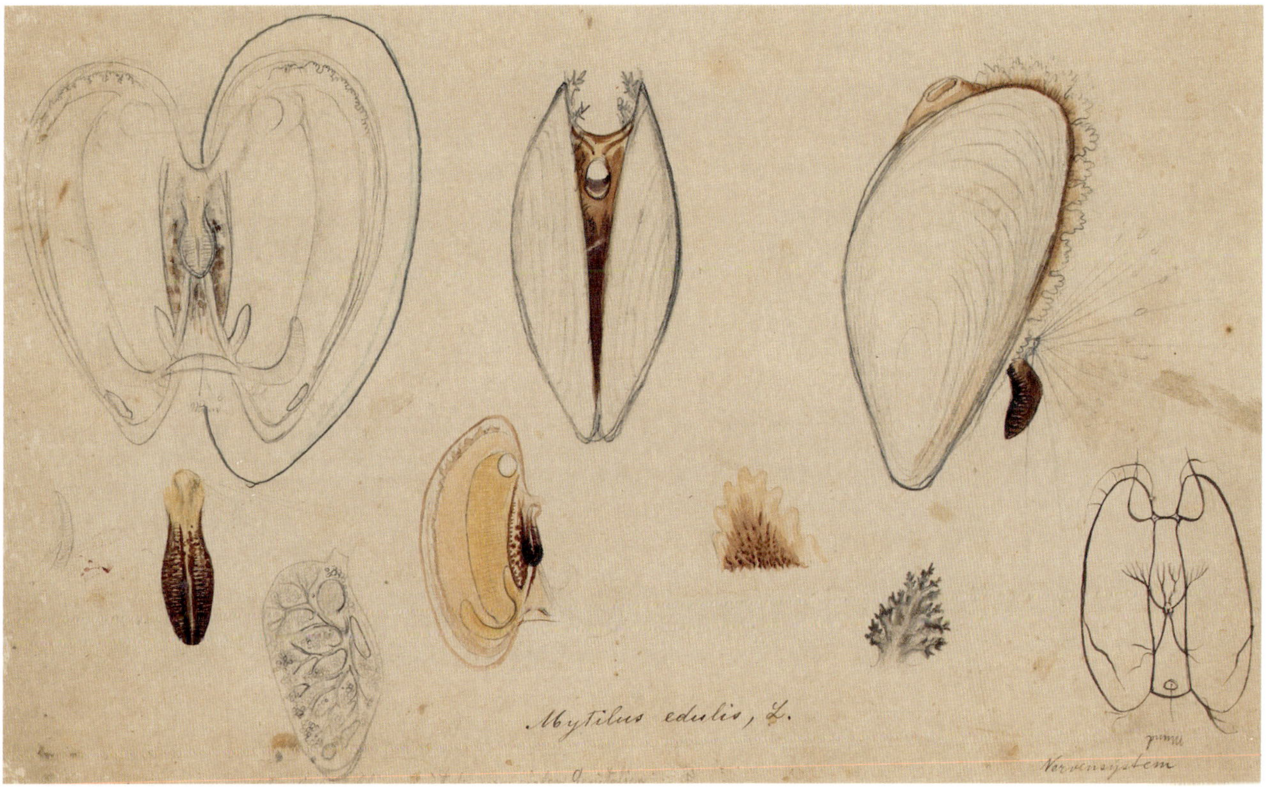
Mytilus edulis, L.
Nervensystem

21.
Philonexis catenulatus
Férussac
Octopus
Order: Octopoda
Model 585
Pencil, ink, and watercolor on paper
29 cm x 24 cm

Octopuses move by crawling along the bottom of the sea, propelled by their suckers. However, when they are frightened, they can move backward rapidly by squirting a jet of water from their siphons (the organs normally used to expel water after respiration). They can also eject an inky substance that clouds the water around them and conceals their escape.

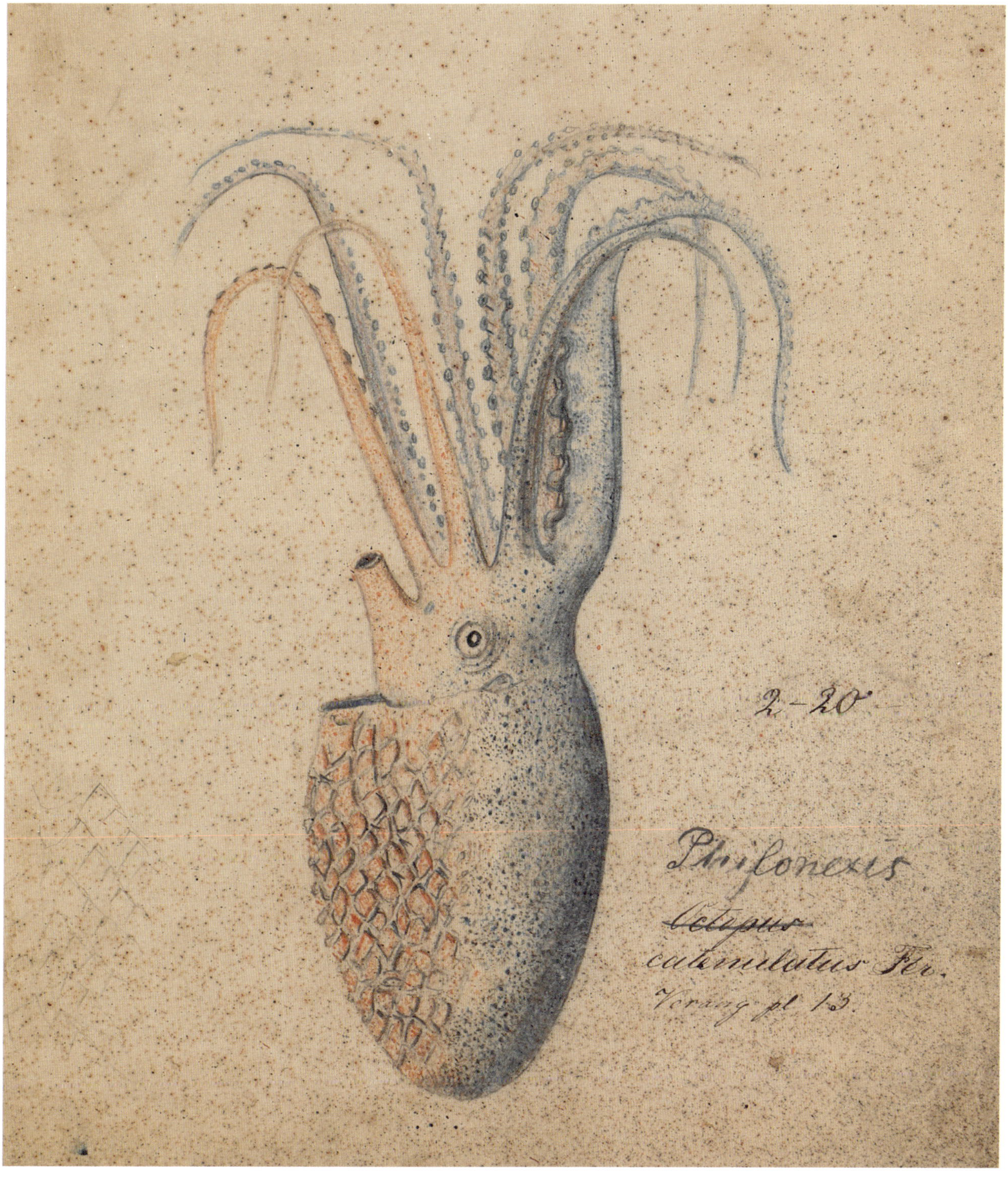
2-20
Philonexis
~~Octopus~~
catenulatus Fer.
Verany pl 13.

22.

Octopus macropus Risso
Octopus
Order: Octopoda
Model 572

Ink and watercolor on paper
39 cm x 65 cm

Octopuses, together with cuttlefish, squids, and flying squids, are cephalopods: the most elaborate mollusks, with sensory organs (such as image-forming eyes) comparable with those of vertebrate animals.

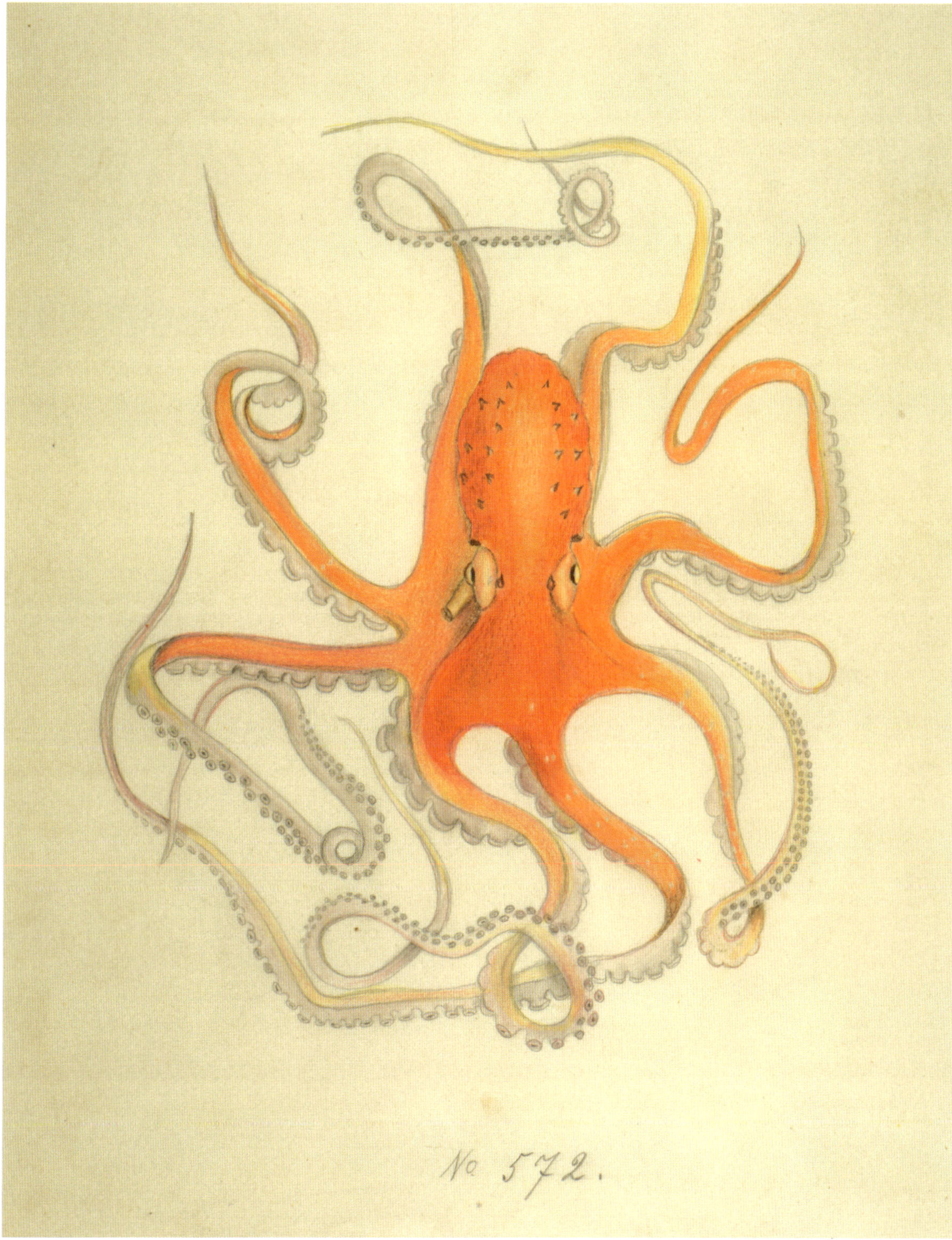
No 572.

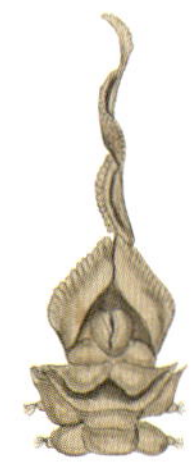

23.
Top:
Serpula contortuplicata
Linnaeus
Model 343
Bottom, left:
Spirographis spallanzanii
Viviani
Model 345
Bottom, right:
Terebella conchilega Pallas
Class: Polychaeta
Pencil, ink, and watercolor on paper
41 cm x 32 cm

The animals illustrated in this drawing are annelids: segmented worms, some 9,000 species of which are found worldwide. These examples belong to the class Polychaeta, or marine worms. Some polychaetes are free-living, while others (like the illustrated specimens) are sedentary, living in soft, flexible tubes attached to the seabed or submerged features such as harbor installations. Most polychaetes reproduce by releasing sperm and eggs into the water. The eggs develop into free-swimming larvae that eventually settle on a suitable surface and mature into adults. The crown of tentacles is used to direct food into the animal's mouth.

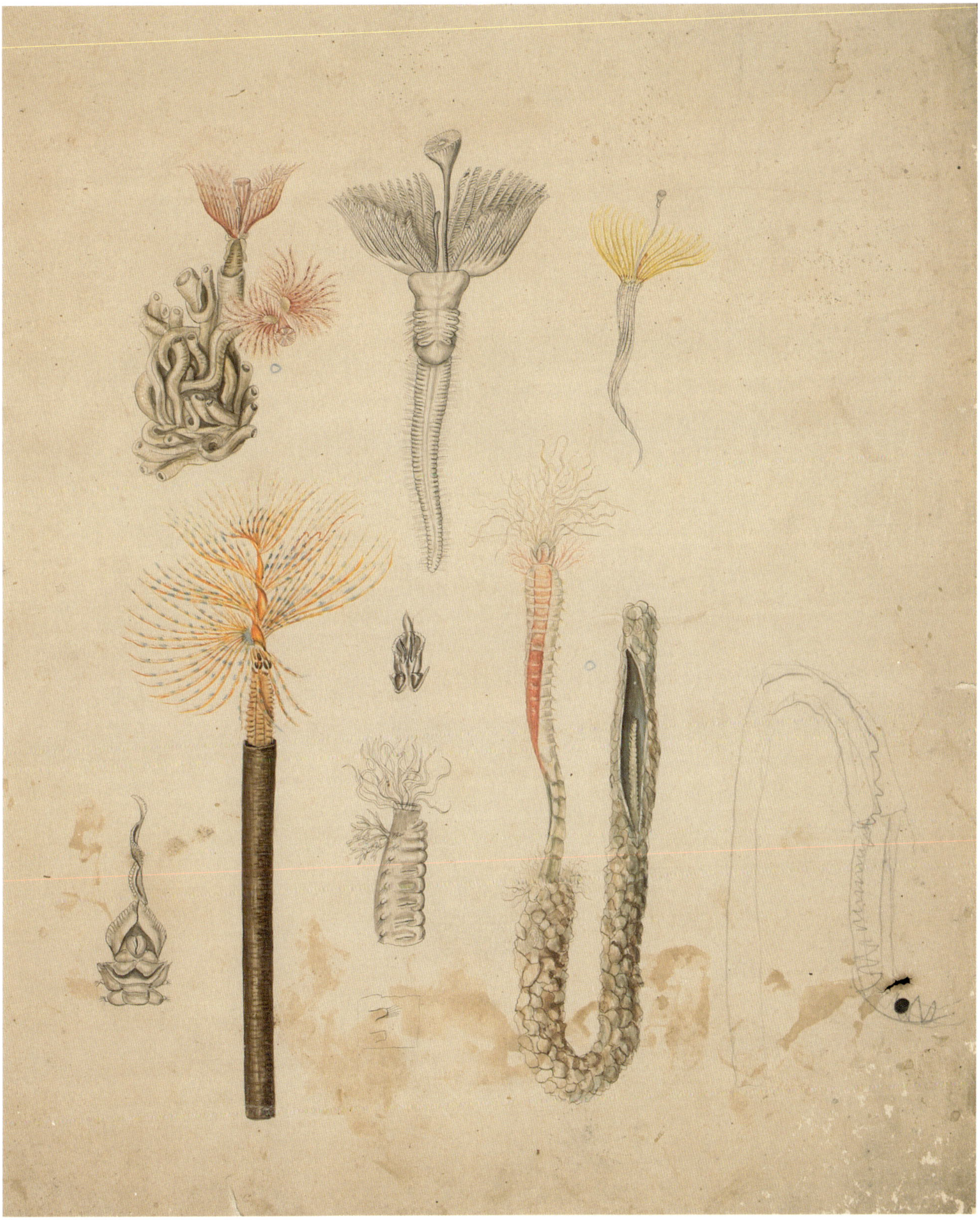

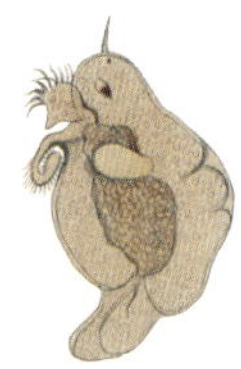

24.
Spirorbis nautiloides Lamarck
(*Spirorbis pagenstecheri*
Quatrefages)
Segmented Worm
Class: Polychaeta
Model 346
Ink and watercolor on paper
23 cm x 33 cm

Animals of the genus *Spirorbis* belong to the phylum Annelida, which comprises more than 9,000 species of segmented worms. The annelids are divided into three classes, of which Polychaeta is the largest, with 6,000 species and a worldwide distribution, usually in marine environments. The other classes are Oligochaeta, earthworms, which live on land, and Hirudinea, leeches, most of which live in fresh water. *Spirorbis* often attaches itself to rocks, but it also lives on seaweed.

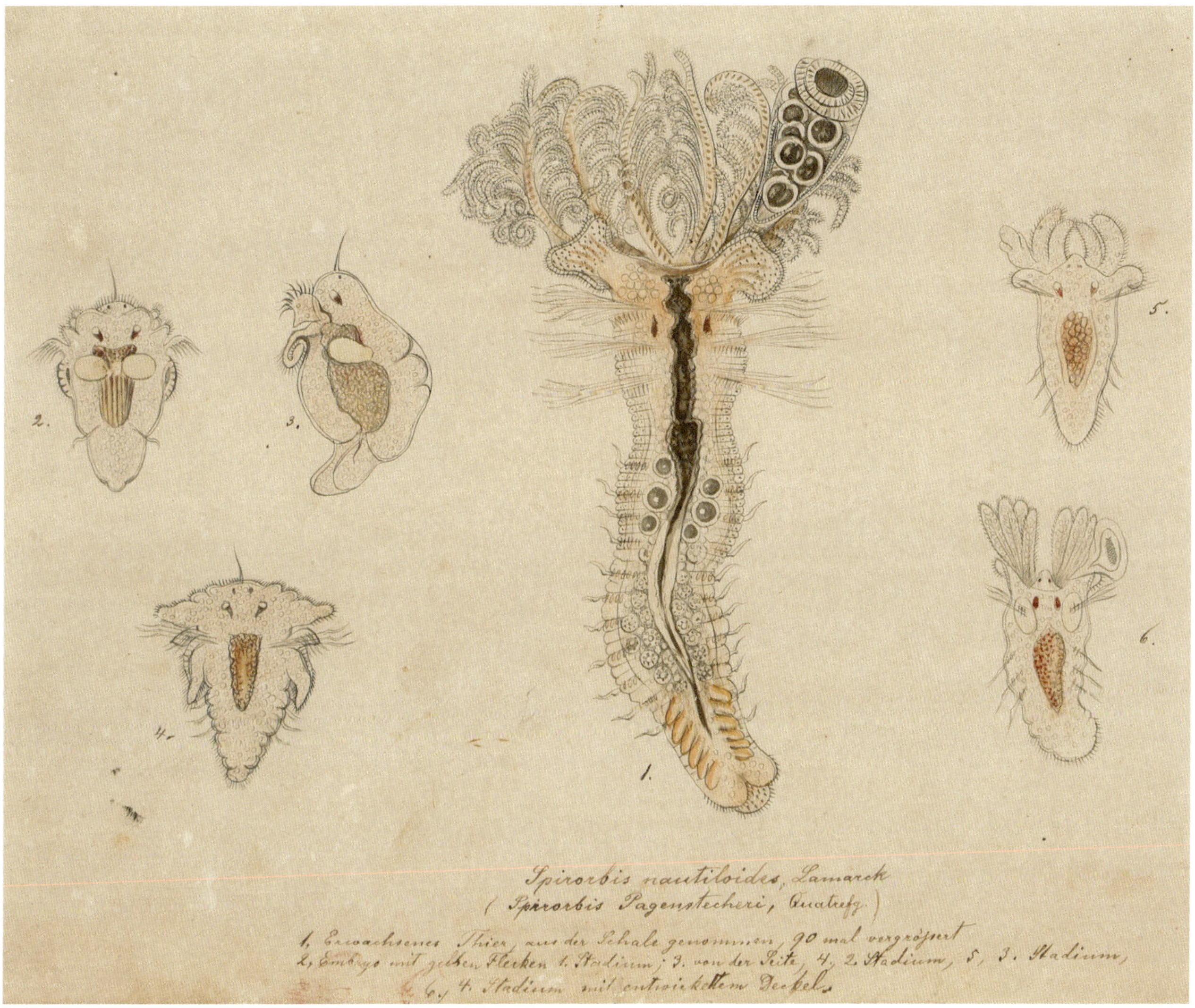
Spirorbis nautiloides, Lamarck
(Spirorbis Pagenstecheri, Quatrefg.)
1, Erwachsenes Thier, aus der Schale genommen, 90 mal vergrößert
2, Embryo mit gelben Flecken 1. Stadium; 3. von der Seite, 4, 2. Stadium, 5, 3. Stadium,
6, 4. Stadium mit entwickeltem Deckel.

25.
Ophiorachna incrassata
Lamarck
Starfish
Class: Asteroidea
Model 258
Pencil, ink, and watercolor on paper
23 cm x 30 cm

Starfishes are common (more than 1,800 species exist today), and they are found in all oceans. Typically, a starfish has five hollow arms containing, on the underside, rows of feet that often terminate in suckers. These feet enable the starfish to move in any direction on the seabed and to cling to steeply inclined surfaces. If a starfish loses an arm, it can grow a new one. Many starfishes are carnivorous, feeding mainly on shellfishes, which they force open with their sucker-tipped arms.

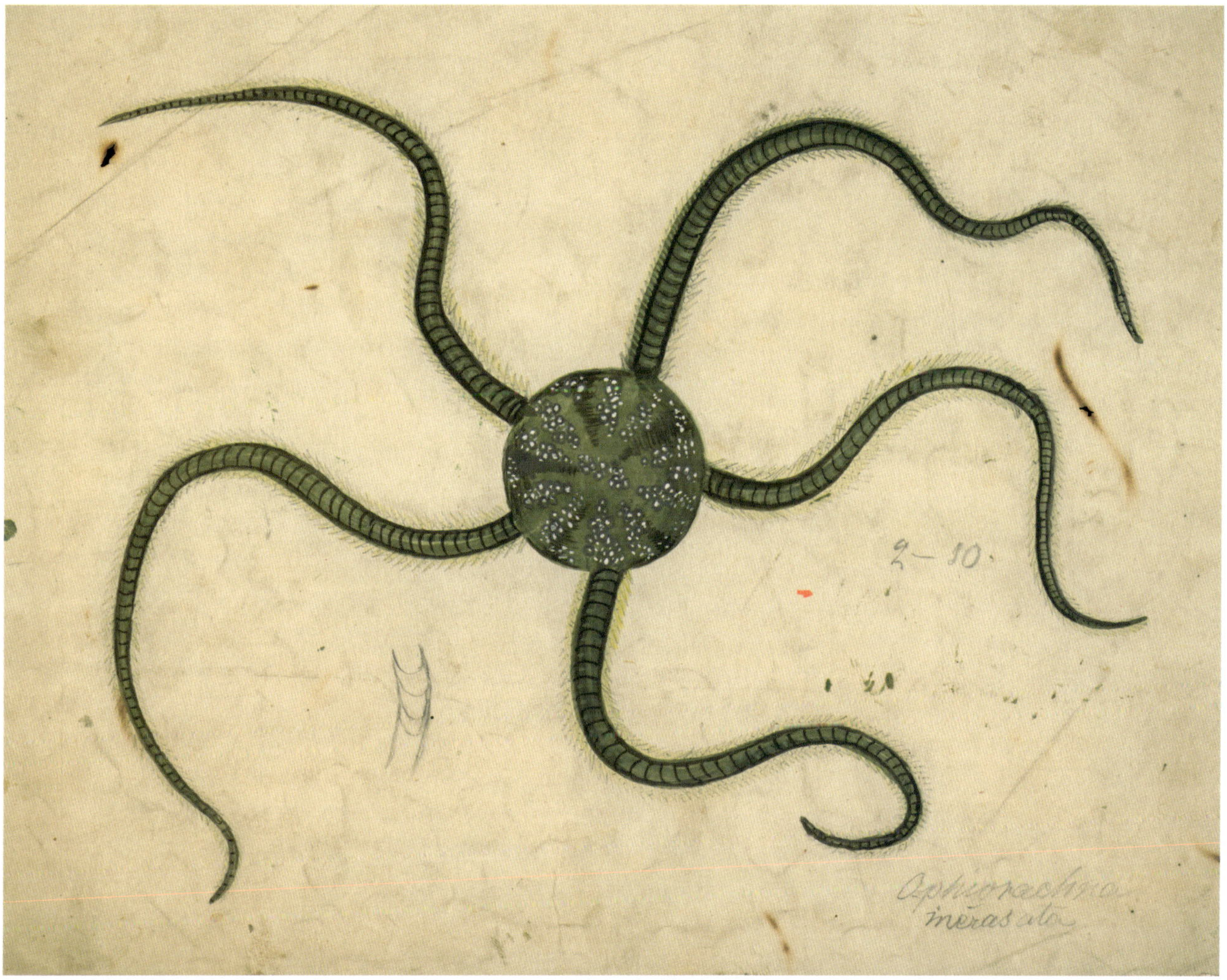
2-10

26.
Oreaster muricatus Linch
Starfish
Class: Asteroidea
Ink and watercolor on paper
24 cm x 29 cm

This starfish was copied from plate III of *Echinodermes peintes d'après nature par les soins de Kühl, Van Hasselt et Sal Müller, membres de la Commission pour l'exploration physique des possessions d'outre-mer de Pays-Bas publiées d'après les cartons du Musée d'histoire naturelle à Leide* (Echinoderms painted after nature by Kühl, Van Hasselt, and Sal Müller, members of the Commission for the Physical Exploration of the Overseas Territories of the Low Countries, published after the plates of the Museum of Natural History of Leiden), ed. J. A. Herklots, Amsterdam: Het Genootschap Natura Artis Magistra, 1869.

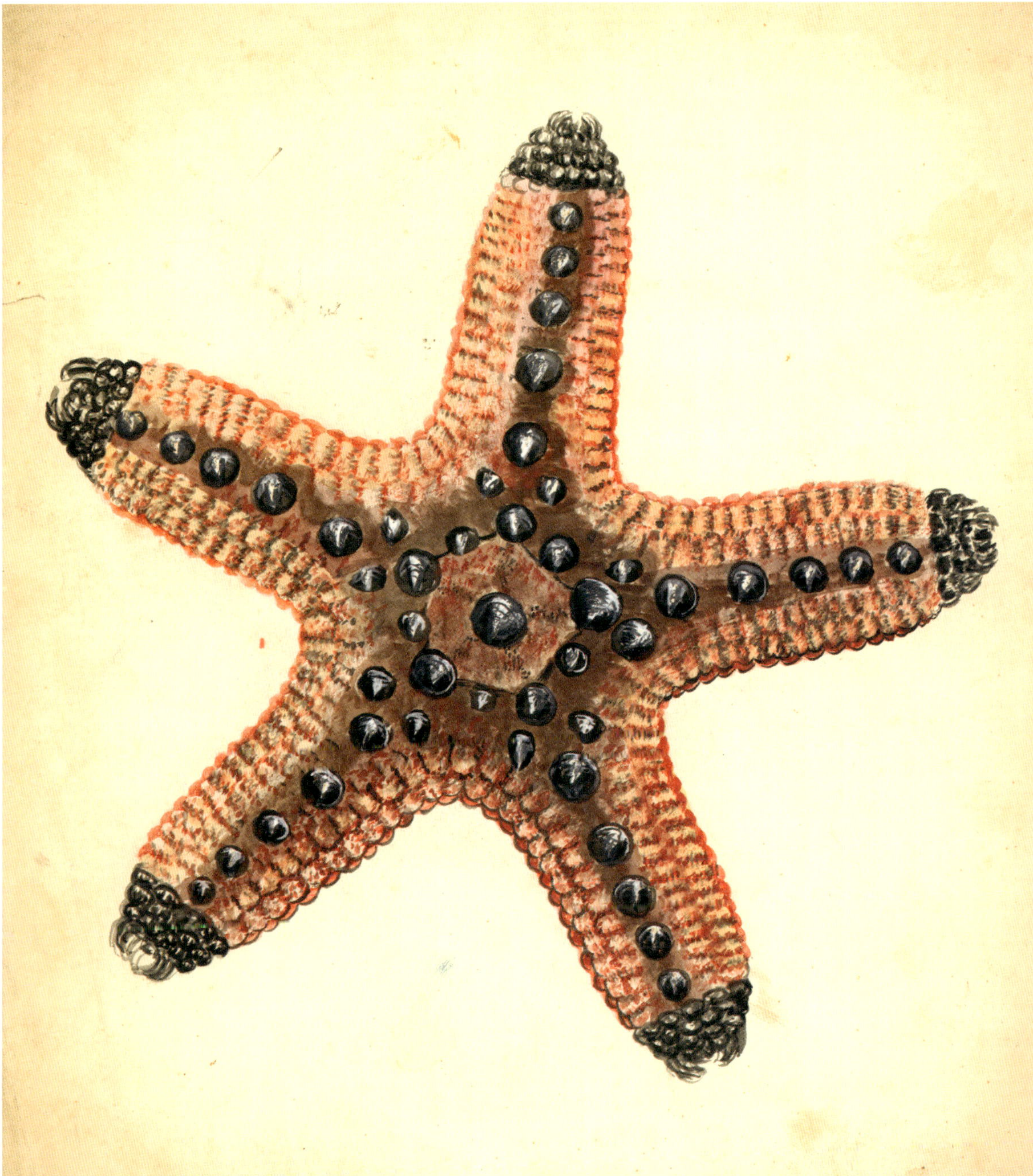

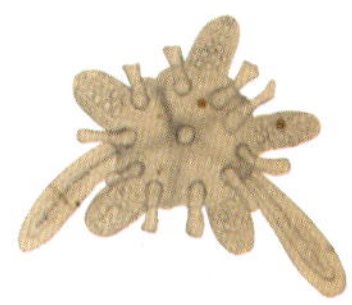

27.
Ophioglypha lacertosa,
Amphiura filiformis Müller
Brittle Star
Class: Ophiuroidea
Model 252
Pencil and ink on paper
25 cm x 29 cm

This detailed and copiously annotated drawing shows 11 stages in the development of a brittle star, from the free-swimming larva with bilateral symmetry to the bottom-dwelling adult with radial symmetry. Starfish, together with sea urchins, sea cucumbers, and sea lilies, belong to the phylum Echinodermata, which includes more than 5,000 species. Brittle stars, which alone number more than 2,000 species, have a worldwide distribution, and they are found at all depths from shallow water to almost four miles below the surface.

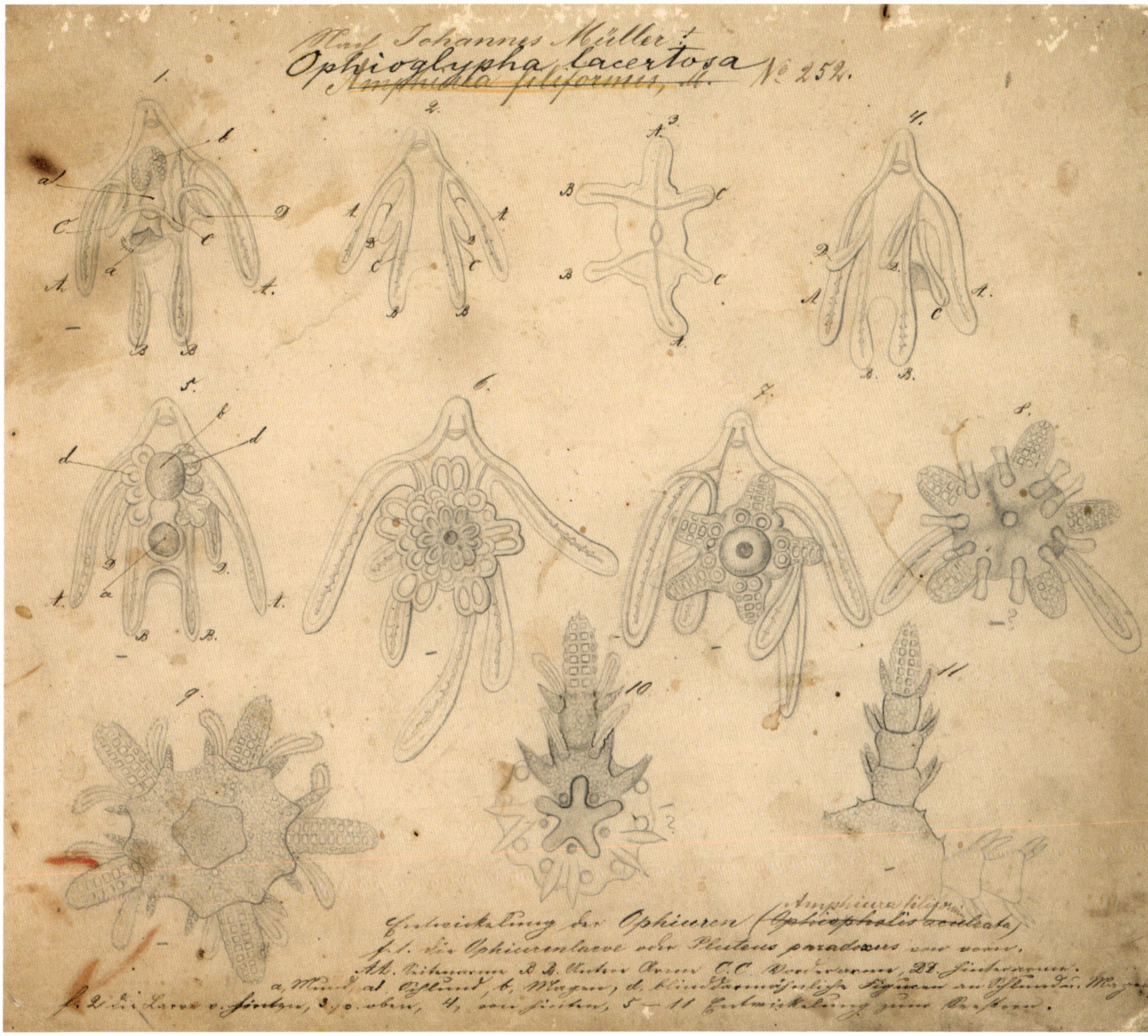
Ophioglypha lacertosa
№ 252.

28.
Sporadipus impatiens Semper
Sea Cucumber
Class: Holothurioidea
Model 279
Pencil, ink, and watercolor on paper
41 cm x 32 cm

Sea cucumbers are marine invertebrates with a soft cylindrical body between two and 200 centimeters in length and a warty skin that resembles a cucumber. Most species have five rows of tube feet and 10 or more tentacles around the mouth. Sea cucumbers are particularly common in the Indian and Pacific Oceans. In the southwestern Pacific, they are boiled, dried, and smoked for use as ingredients in soups. In New Guinea, the French name for the sea cucumber, *bêche-de-mer*, has been anglicized to "Beach-la-Mar" and is used to denote the local version of pidgin English.

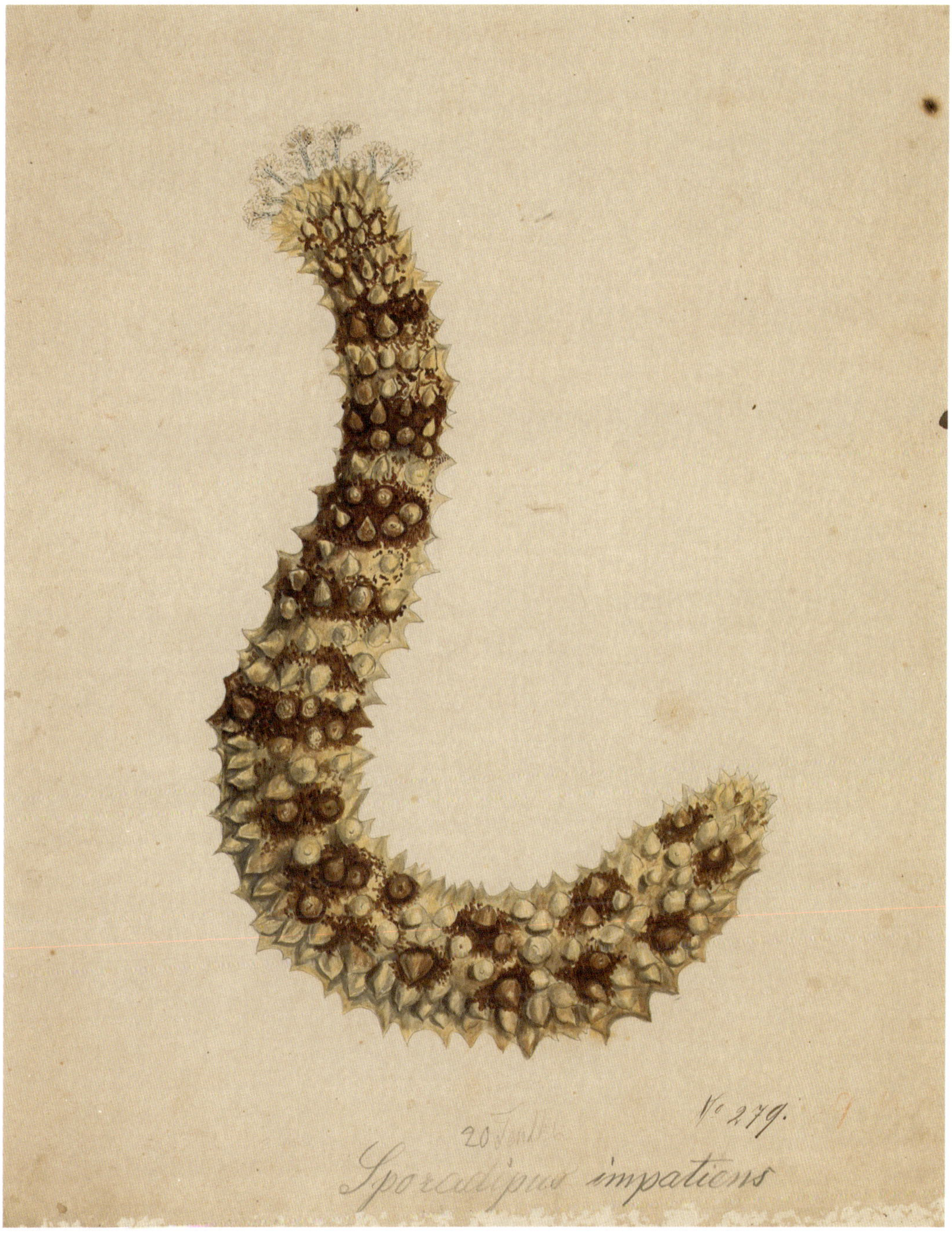
N° 279.
Sporadipus impatiens

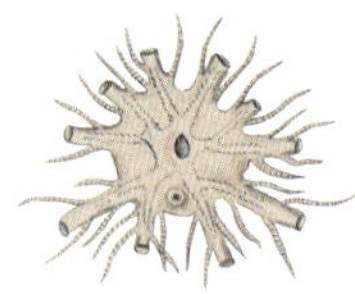

29.
Comulata mediterranea
Lamarck
Sea Lily
Class: Crinoidea
Model 250
Ink and watercolor on paper
41 cm x 32 cm

Sea lilies, like starfish and sea cucumbers, belong to the phylum Echinodermata, which contains more than 5,000 species of marine invertebrates. Crinoids live on the bottom of the sea. They have a stalk with, at the top, a cup with a central mouth. Five arms with branches extend upward from the edges of the cup. Grooves run down the arms and the branches, and these are lined with hairlike cilia. Food falling from above is directed by the cilia down the grooves and into the animal's mouth.

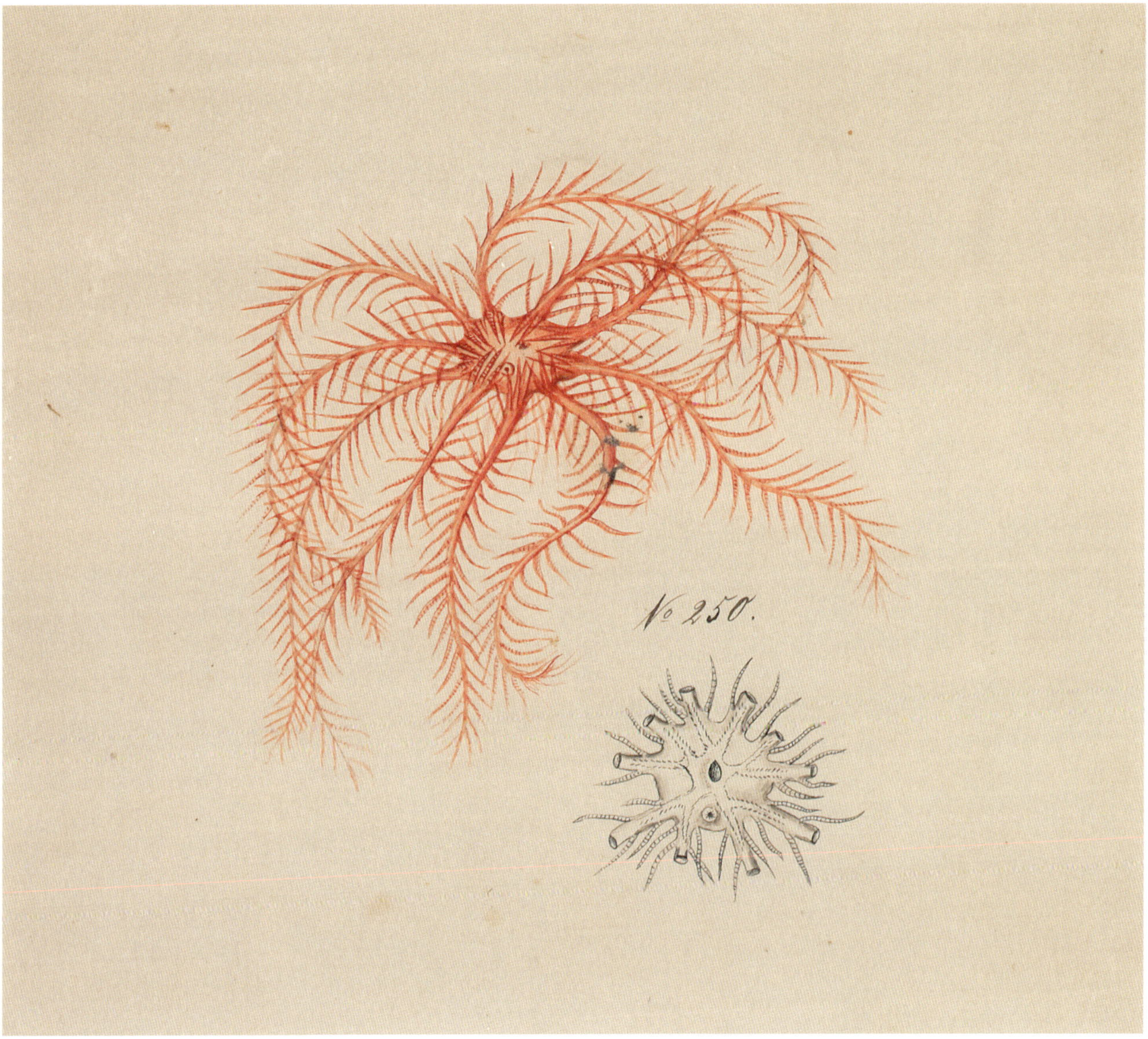
No 250.

30.
Doliolum mülleri Krohn
Sea Squirt
Family: Doliolidae
Model 621

Ink and watercolor on paper
25 cm x 33 cm

Sea squirts and their relatives are small animals that are found in every ocean. Indeed, more than 2,000 species exist worldwide. Some sea squirts are sessile (that is, they are permanently attached to the seabed), and others are free-swimming. They range in size from a few millimeters to 30 centimeters, and the floating species may form colonies that are one or more meters across. Most sea squirts have two apertures: one to take in water, which contains food and oxygen, and the other to expel the water after it has been filtered.

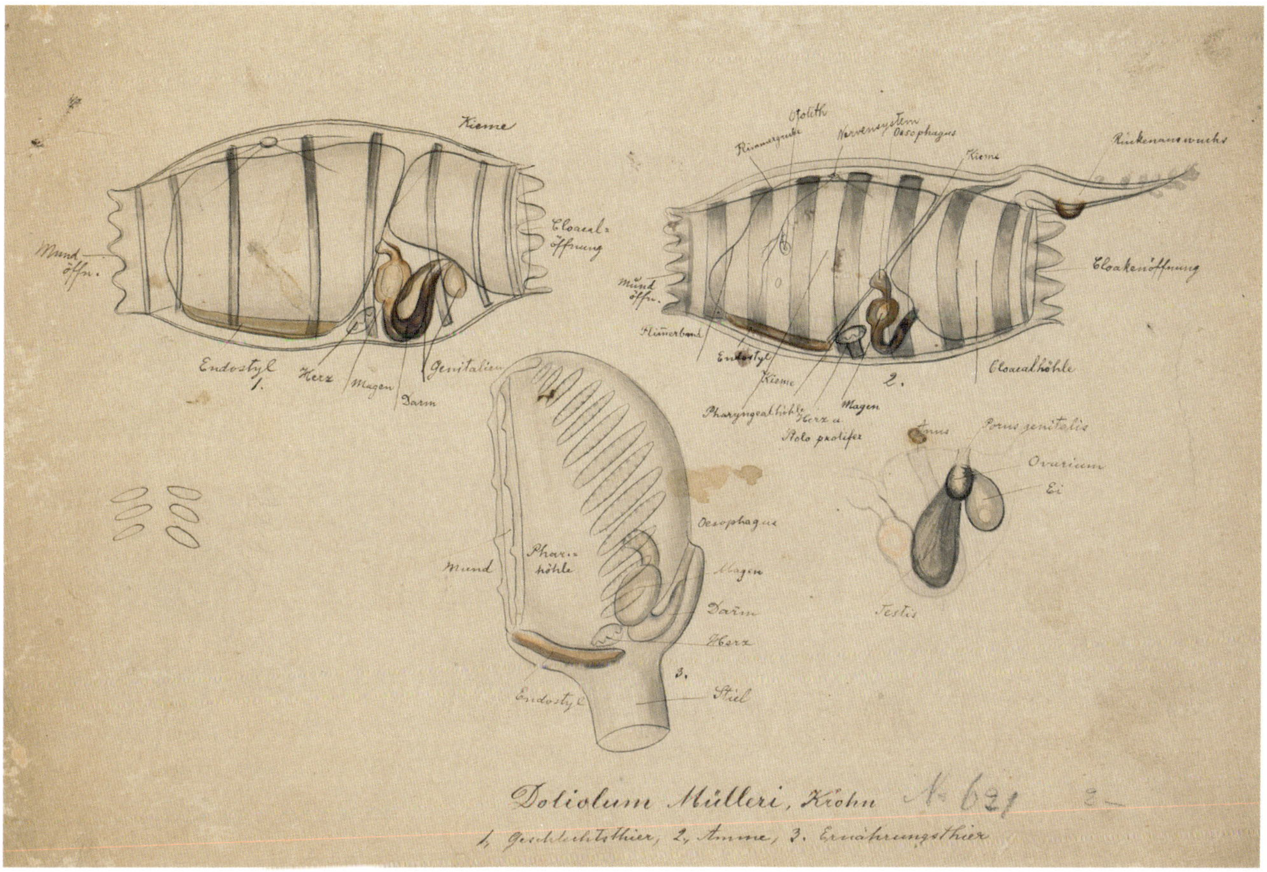
Kieme
Mund öffn.
Cloacal-öffnung
Endostyl
1.
Herz
Magen
Darm
Genitalien
Otolith
Nervensystem
Oesophagus
Kieme
Rückenauswuchs
Mund öffn.
Cloakenöffnung
Flimmerband
Endostyl
Kieme
2.
Cloacalhöhle
Pharyngealhöhle
Herz u.
Stolo prolifer
Magen
Anus
Porus genitalis
Ovarium
Ei
Testis
Oesophagus
Mund
Pharynxhöhle
Magen
Darm
Herz
3.
Endostyl
Stiel
Doliolum Mülleri, Krohn
1, Geschlechtsthier, 2, Amme, 3. Ernährungsthier

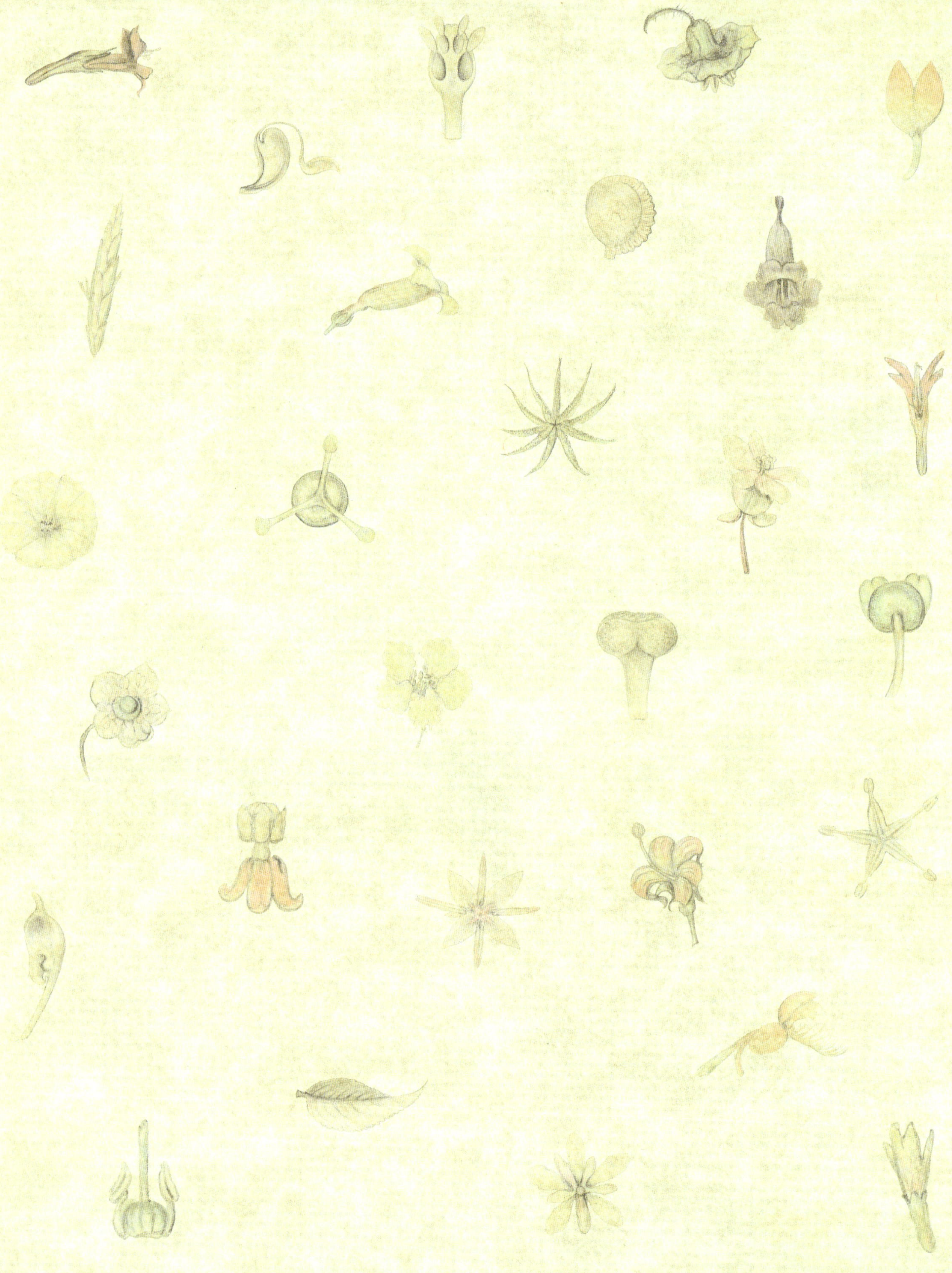

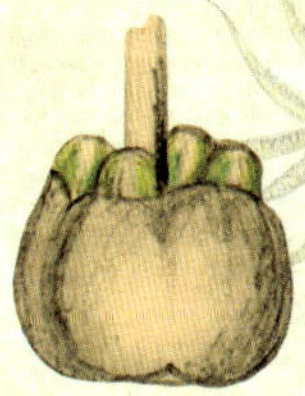

Drawings of Plants

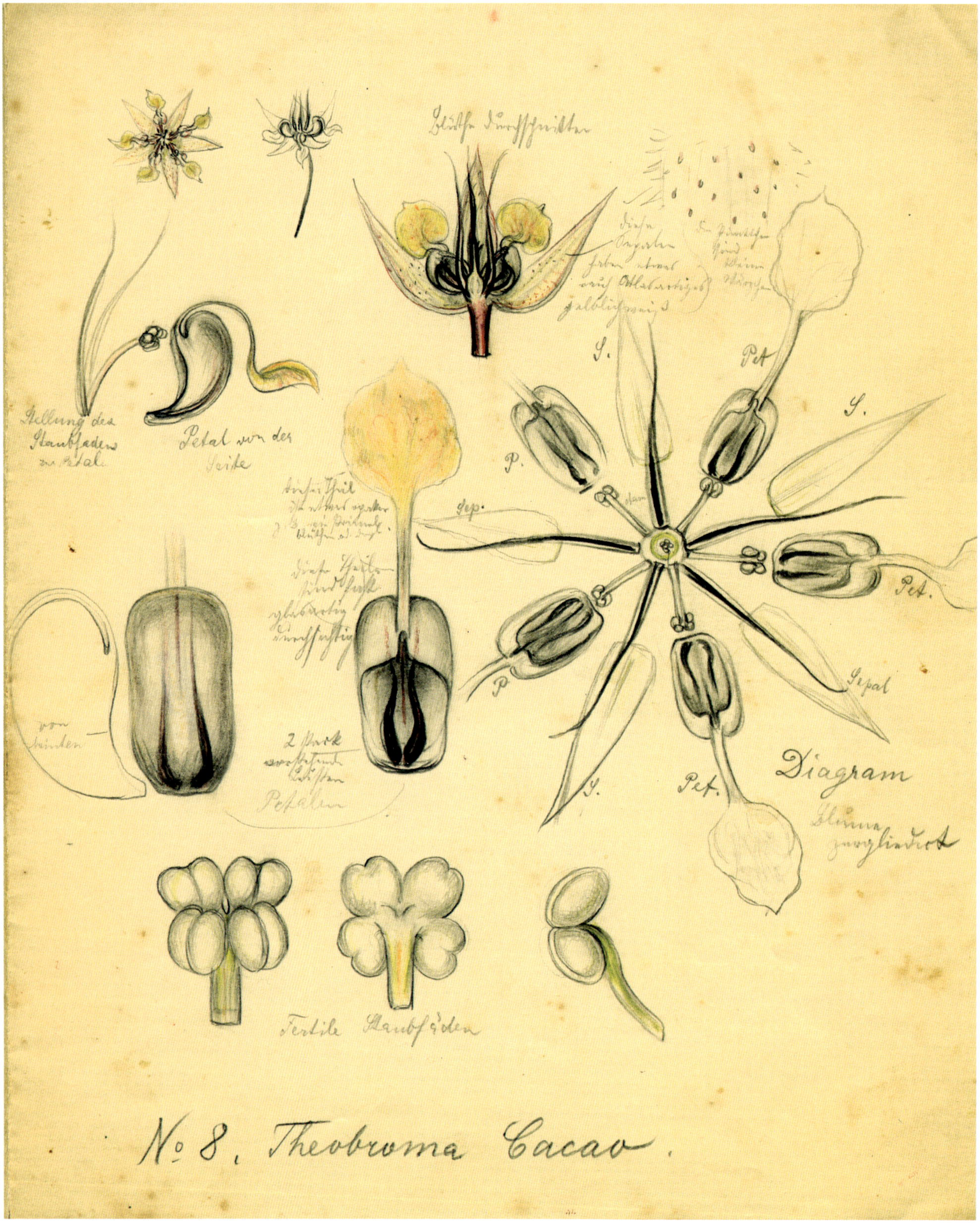
Petal von der Seite
Diagram
Pet.
Sepal
Fertile Staubfäden
No 8. Theobroma Cacao.

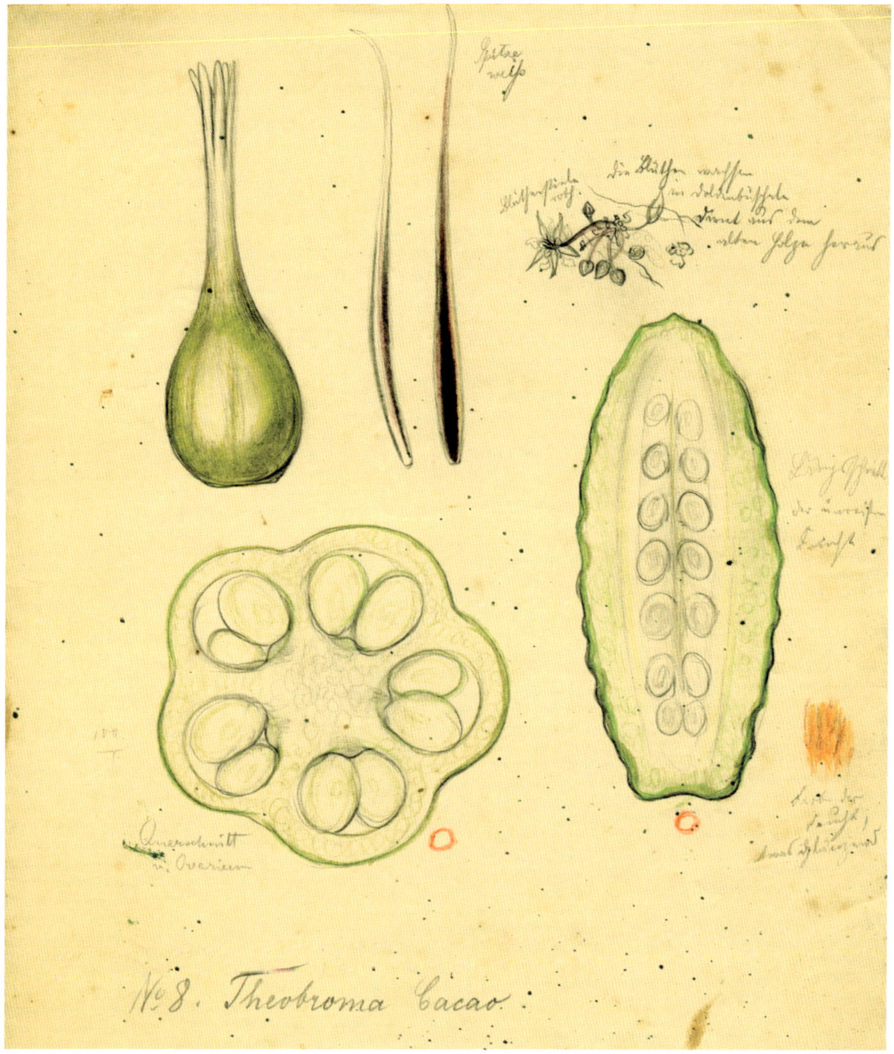

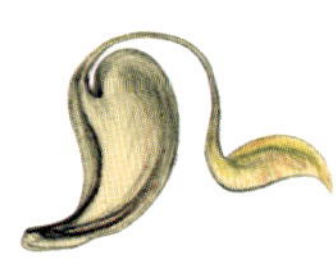

1.
Theobroma cacao L.
Cacao
Sterculiaceae
Model 374 (1893)
L. and R. Blaschka
Drawing no. 8
Colored pencil on paper
21 cm x 26.7 cm

Chocolate is made from the seeds of this tropical American native (*above*). Rudolf Blaschka studied the complex flowers, which grow on short stalks from the tree trunk (*opposite*). This 1892 drawing, made in Jamaica, illustrates the relationship of the various parts. In "diagram of the dissected flower" (*opposite, right*), "S" and "P" stand for "sepals" and "petals" respectively. The enlarged petals (*opposite, left*) highlight the "2 strong protruding ribs," with the cup "almost glasslike in transparency."

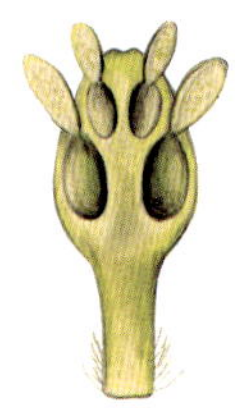

2.
Cinnamomum "officinale"
(currently *C. verum* J. Presl)
Cinnamon
Lauraceae
Model 447 (1894)
L. and R. Blaschka
Drawing no. 19
Colored pencil on paper
21 cm x 26.7 cm

Quills of cinnamon are harvested from the bark of the evergreen tree shoots. Rudolf Blaschka, who was intrigued by the spice plants he observed in Jamaica, wrote to his parents on March 6, 1892, "Today I had the pleasure of seeing the flowers of the cinnamon tree, *Cinnamomum officinale* [*sic*], which grows almost wild here." The drawing focuses on the flaplike valves that release pollen from two whorls of fertile stamens "a" and "b"; the third whorl, "c," is sterile.

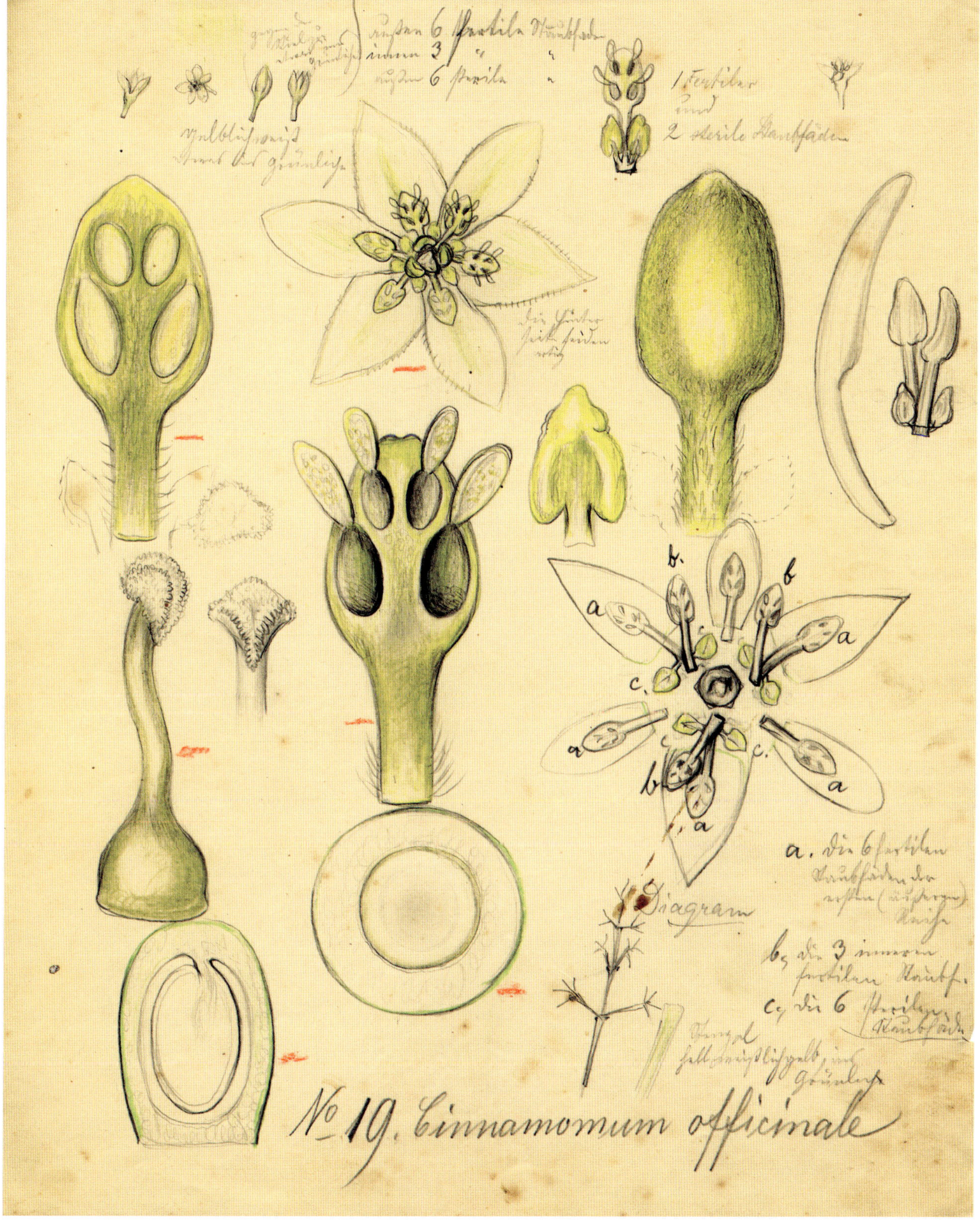
№ 19. Cinnamomum officinale
Diagram

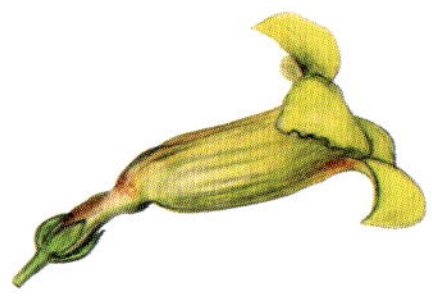

3.
Allamanda cathartica L.
Golden Trumpet
Apocynaceae
Model 377 (1893)
L. and R. Blaschka
Drawing no. 20
Colored pencil on paper
21 cm x 26.7 cm

This colorful tropical vine is widely cultivated as an ornamental. Rudolf Blaschka collected specimens near Kingston, Jamaica, in 1892. The drawing delineates the overlapping "finely wrinkled, shiny" petal lobes as they appear in the bud stage, and fully opened in a tubular blossom that is "matte, fiery, nice yellow somewhat frosted gold." In the circle below the flattened view, Blaschka even sketched the bunches of white hairs around the anthers at the base of the flower.

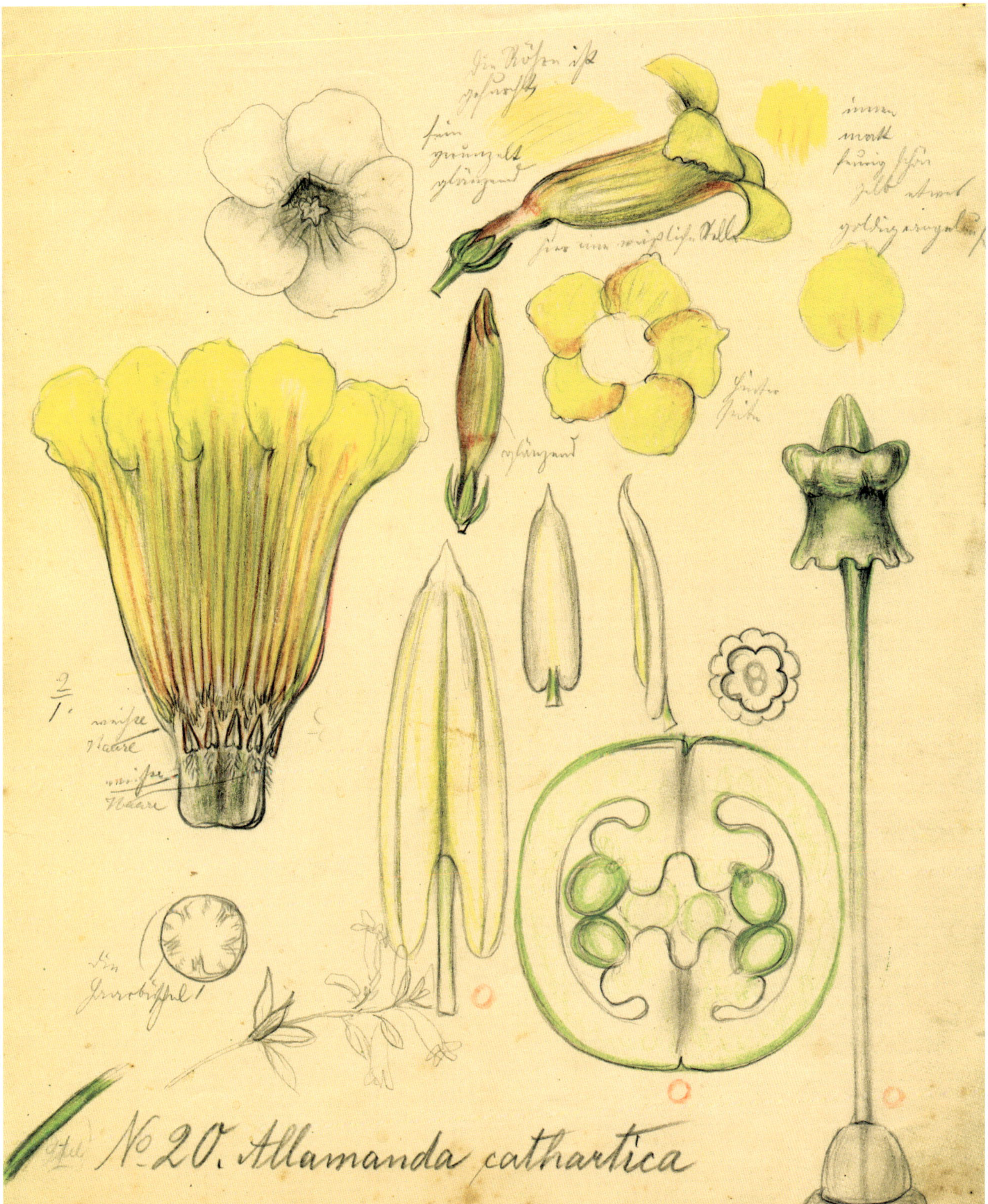
No 20. Allamanda cathartica

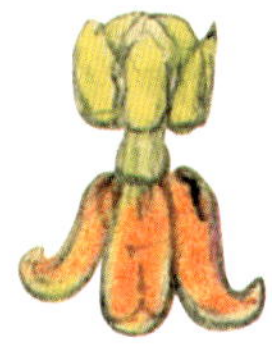

4.
Asclepias curassavica L.
Blood Flower Milkweed
Asclepiadaceae
Model 526 (1895)
L. and R. Blaschka
Drawing no. 22
Colored pencil on paper
21 cm x 26.7 cm

A sketch above the enlarged flower diagrams the central placement of the gynoecium, a collective of pistils, which is receptive to pollen (*detailed at middle right*). Surrounding it are five orange petals fused into corona hoods and horns (*upper right*). Pairs of pollen sacs (pollinia; *left*) sit inside adjacent corona hoods. The arms are joined by an adhesive gland, which attaches to the visiting insect that carries it off, providing pollen to other flowers.

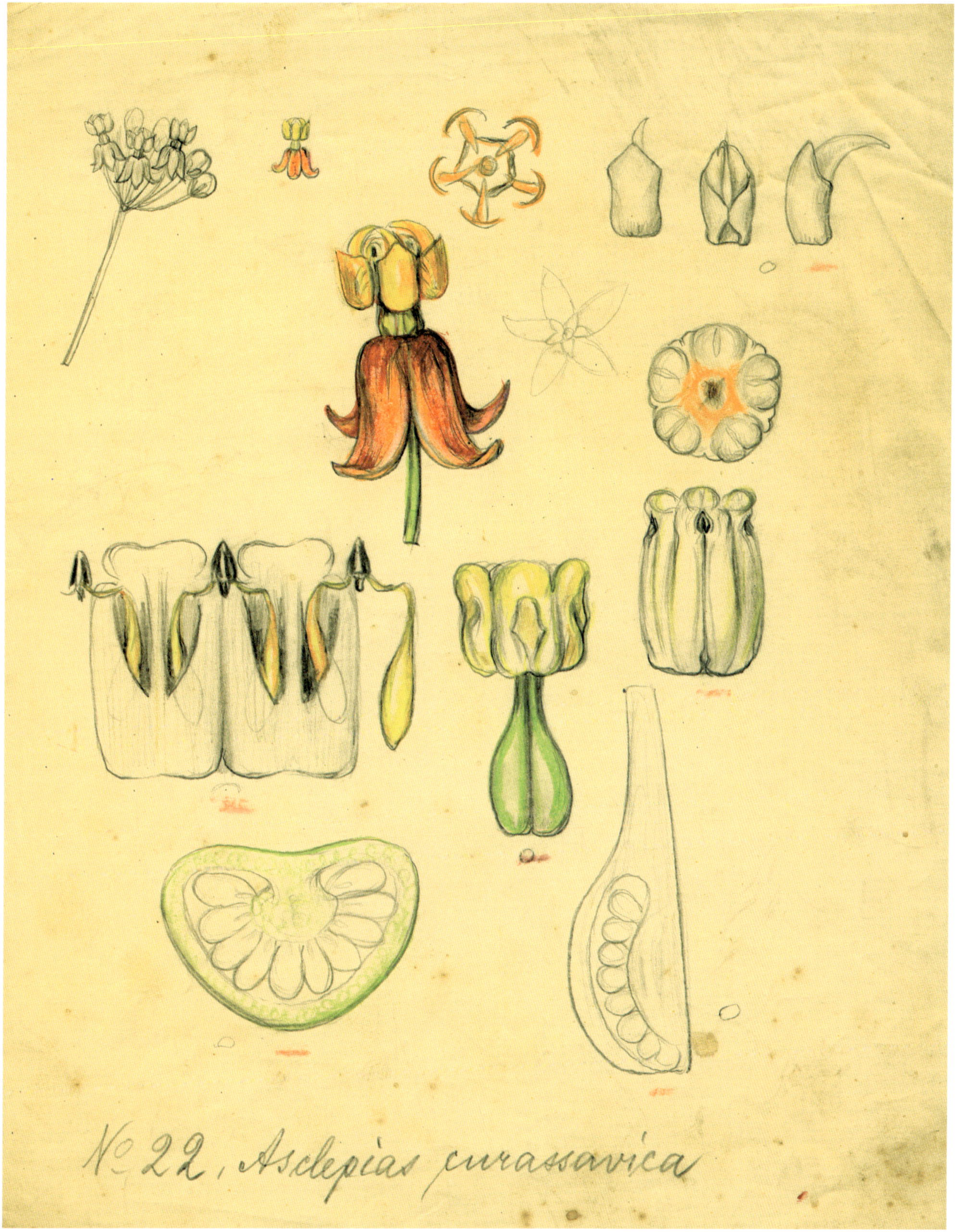
№ 22. Asclepias curassavica

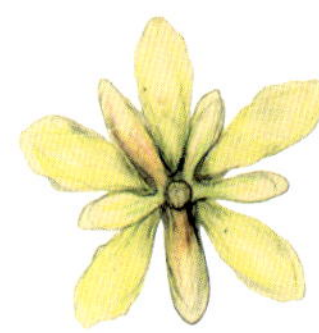

5.
Haematoxylon campechianum
(*Haematoxylum*
campechianum L.)
Bloodwood Tree
Fabaceae
Model 480 (1894)
L. and R. Blaschka
Drawing no. 24
Colored pencil on paper
21 cm x 26.7 cm

Although the bloodwood is considered to be an invasive plant today, its heartwood (darker interior part) has been an important source of tannins and of the purplish red dye used in histology. The clustered trunks of this small, thorny, leguminaceous tree have a corrugated appearance. Rudolf Blaschka's 1892 illustration from Jamaica skillfully depicts the flower in various stages (*top*), recording the color as "primrose yellow," with additional details for placement of the anthers. The branches, he wrote, are "green like *Prunus serotina*" (black cherry).

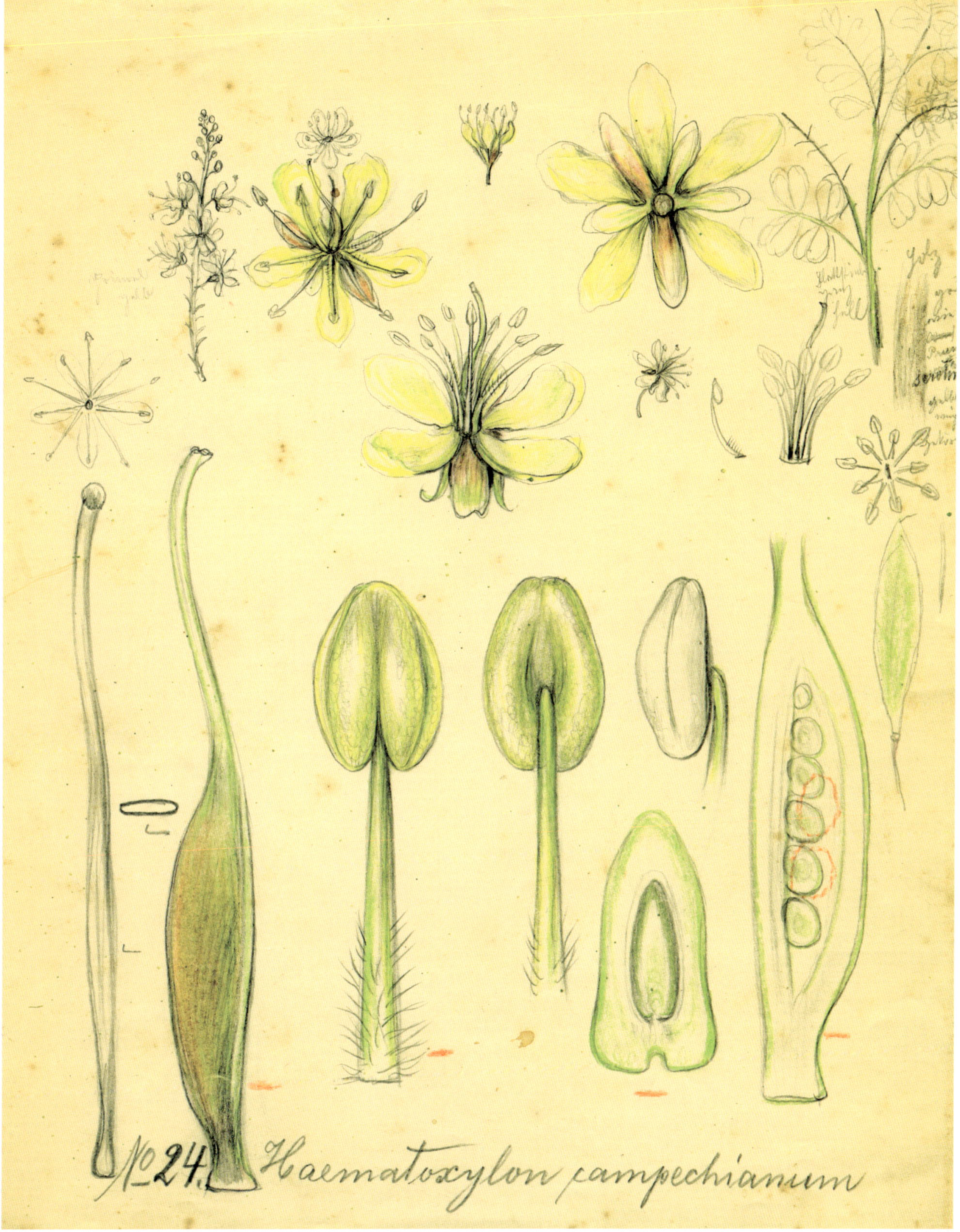
№ 24. Haematoxylon campechianum

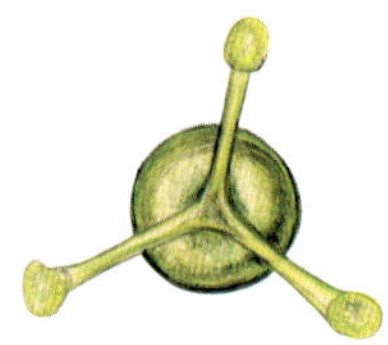

6.
Erythroxylon coca
(*Erythroxylum coca* Lam.)
Coca Plant
Erythroxylaceae
Model 449 (1894)
L. and R. Blaschka
Drawing no. 27
Colored pencil on paper
21 cm x 26.7 cm

In 1892, Rudolf Blaschka drew this important medicinal plant, which is native to the Andes. Less than a decade earlier, pharmaceutical companies had hired botanists to find sources of the plant. The narcotic is extracted from the leaves (*lower right*). The drawings of the flower (*top right*) depict the circle of mature anthers that later relax against the petals as the three-part pistil (*in the middle*) becomes receptive to pollen. The rough petals (*upper left*), Blaschka wrote, have "skin outgrowths."

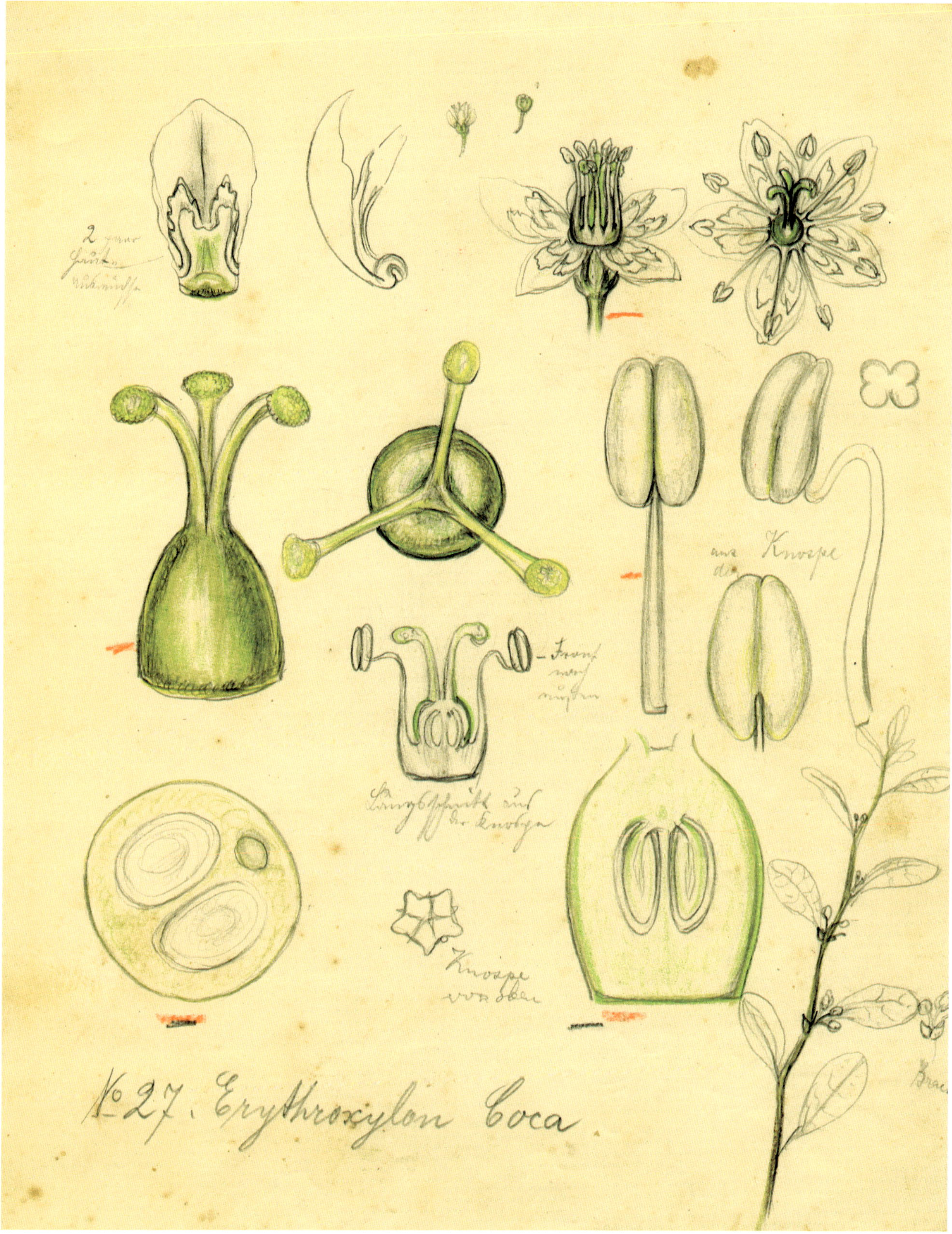
Knospe
No 27. Erythroxylon Coca

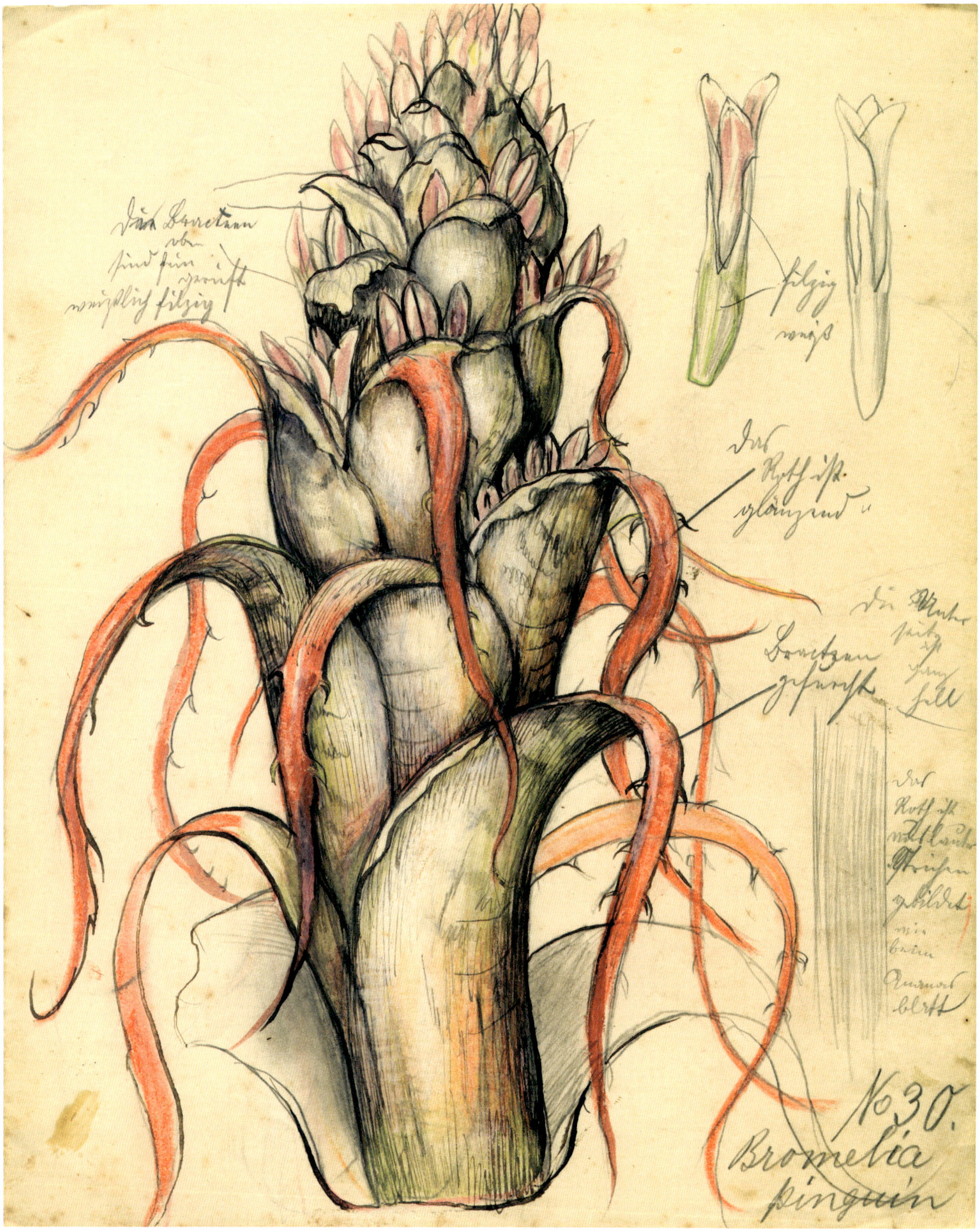
No 30.
Bromelia
pinguin

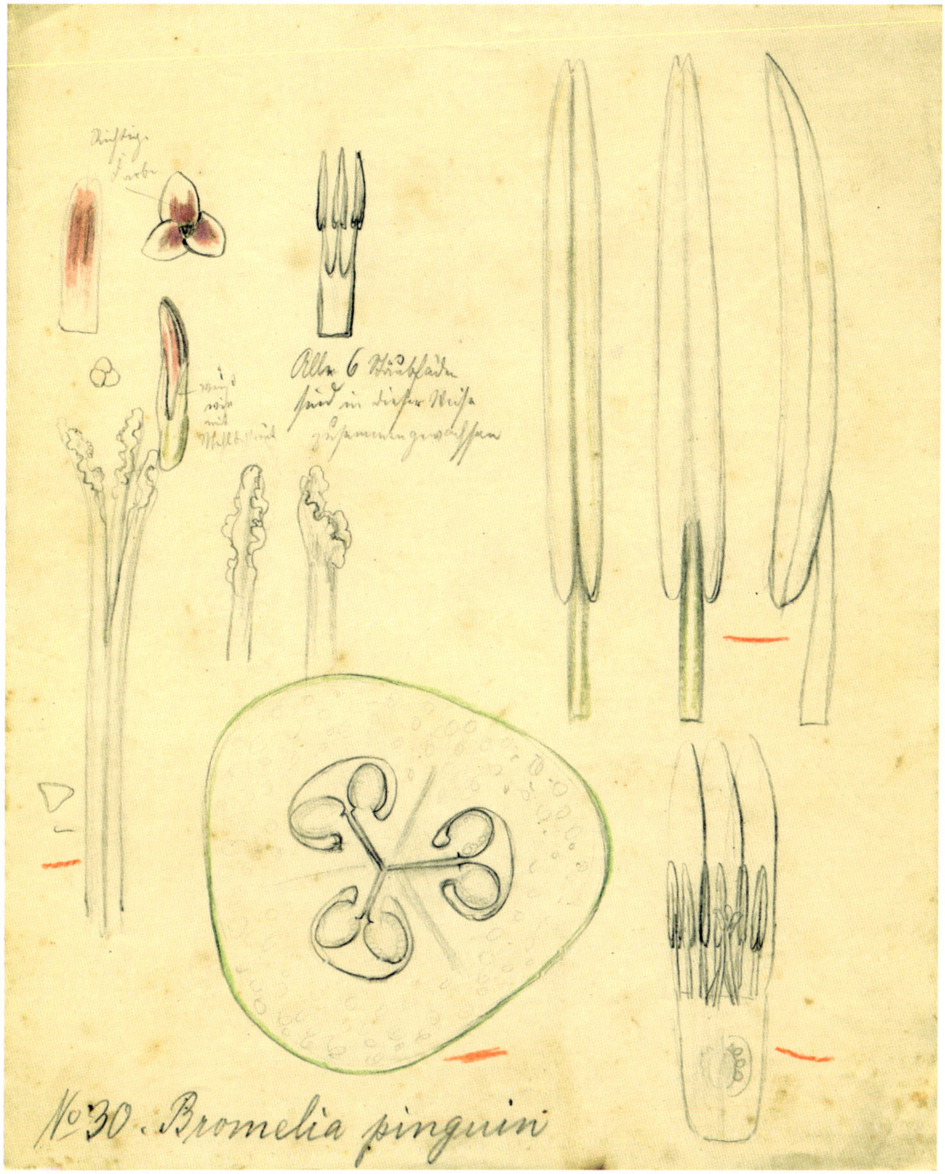

7.
Bromelia pinguin L.
Rat Pineapple
Bromeliaceae
Model 151 (1894)
L. and R. Blaschka
Drawing no. 30
Colored pencil and ink on paper
21 cm x 26.7 cm

This edible pineapple is a perennial monocot that is found throughout tropical America. Rudolf Blaschka observed it in Jamaica a year after he and his father had reproduced models of the larger commercial species, *Ananas comosus*. The drawing indicates the "white felt" texture of the floral bracts, which are shown in the drawings to the right (*opposite*). The stiff, spiny leaves characteristic of bromeliads are "furrowed" and "shiny." "The red is composed of many small stripes like the pineapple leaf," Blaschka wrote.

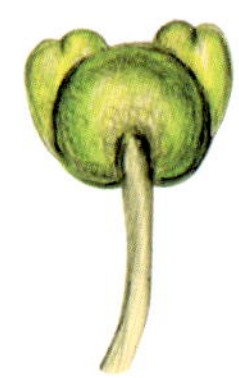

8.
Inga vera Willd.
River Koko
Fabaceae
Model 556 (1895)
L. and R. Blaschka
Drawing no. 32
Colored pencil on paper
21 cm x 26.7 cm

River koko is a tree that is commonly found along riverbanks. The petals of its flowers are fused into a hairy corolla tube that surrounds an ovary with a bright red style (*right*), and the frilly stamens are fused into a tube at the base (*cutaway at right*). Enlargements of the green-colored anthers, attached to a fleshy disk, are shown in several views (*lower left*). The fruit (*lower right*) is a legume.

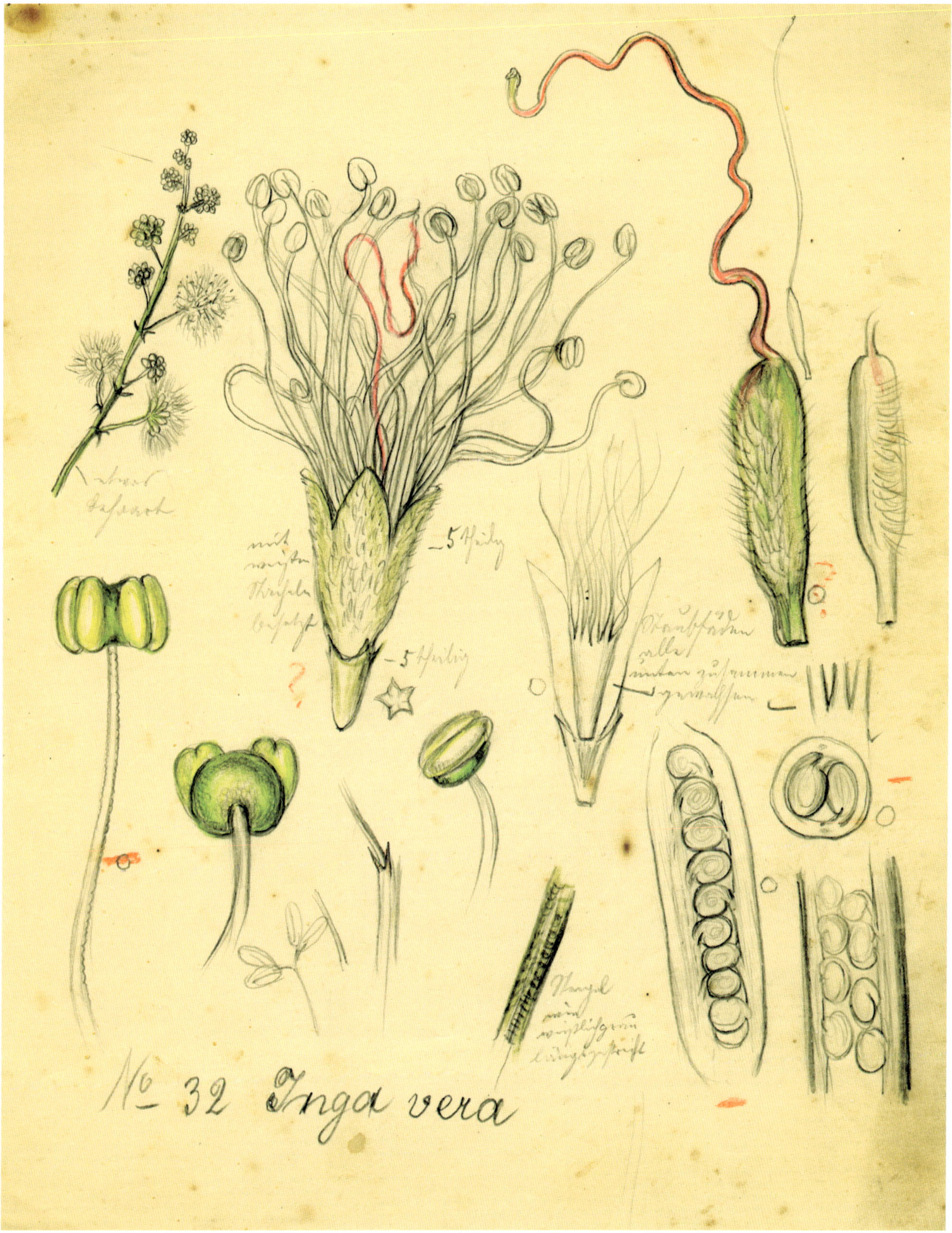
No 32 Inga vera

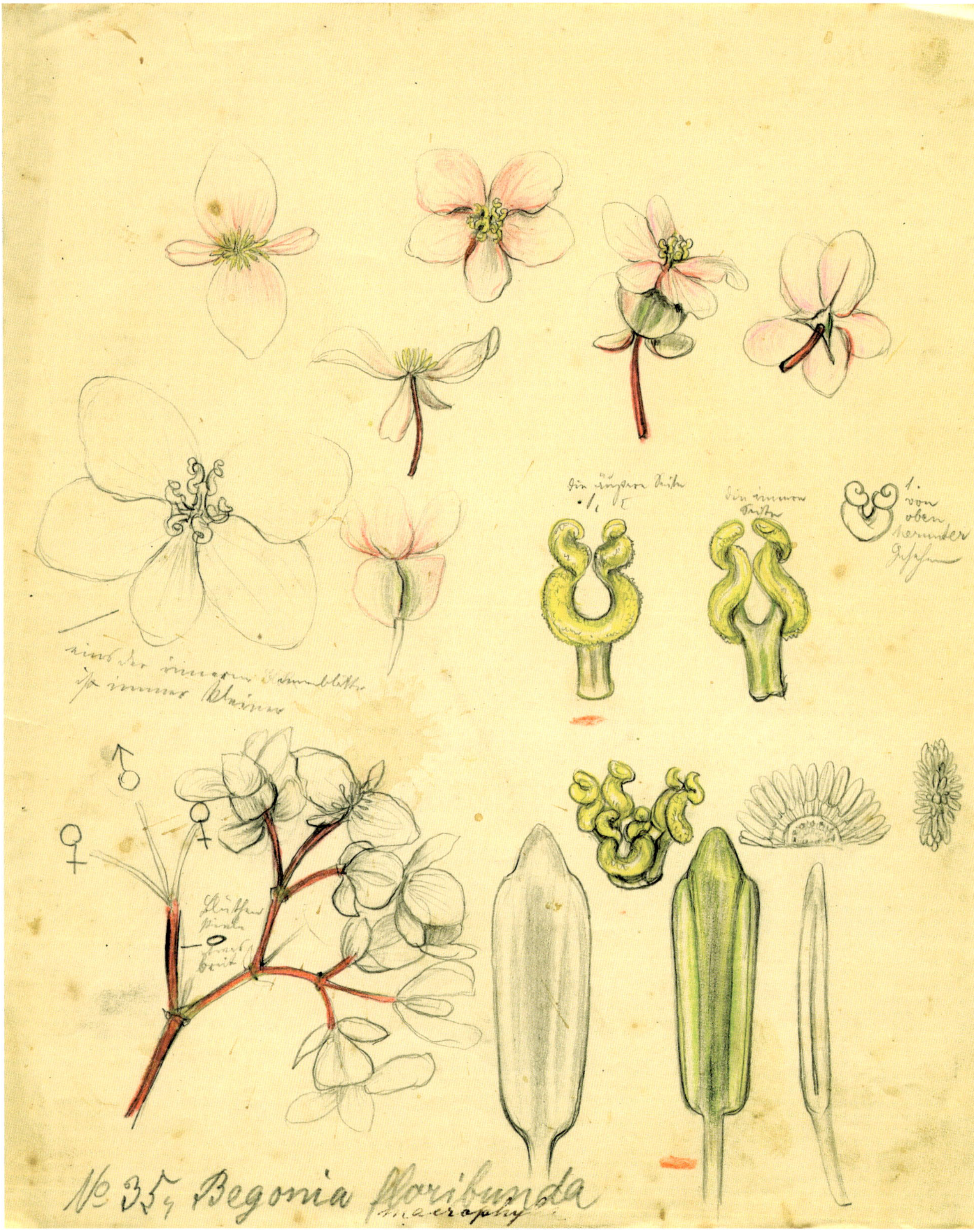

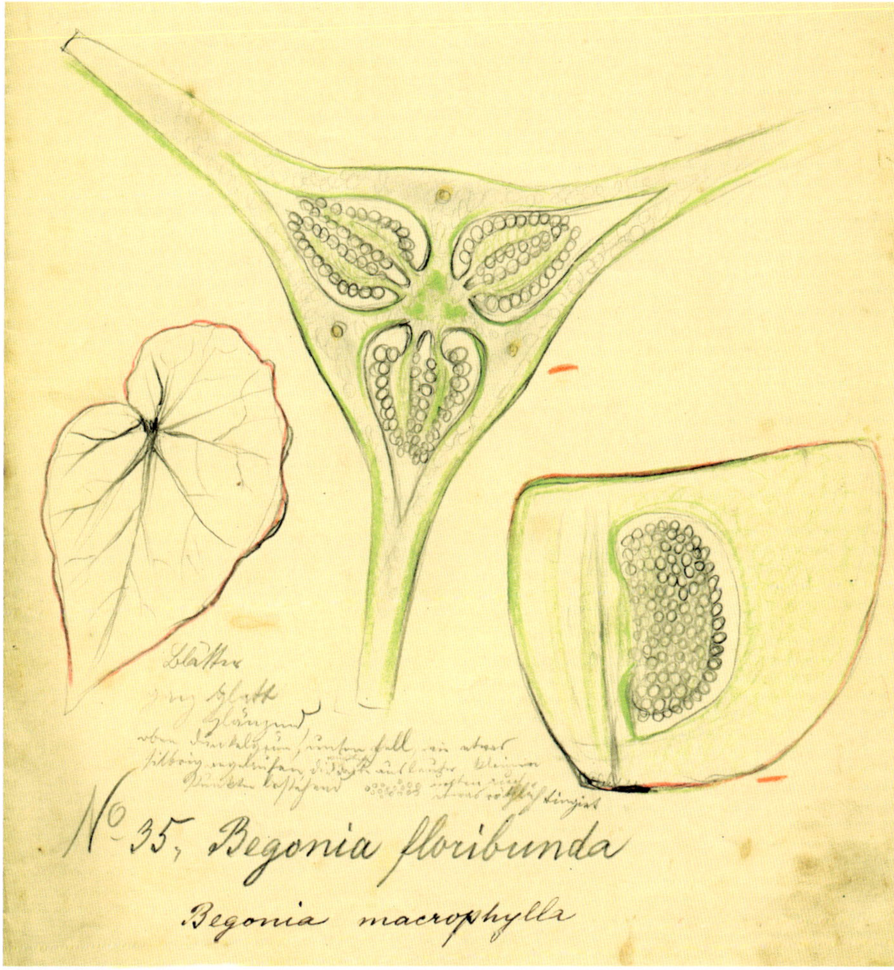

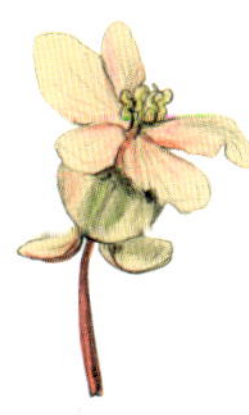

9.
Begonia floribunda Carrière
Begonia
Begoniaceae
Model 562 (1895)
L. and R. Blaschka
Drawing no. 35
Colored pencil on paper
21 cm x 26.7 cm

On the way to Castleton Gardens, Jamaica, Rudolf Blaschka saw what he called "impressive forests with indescribably beautiful vegetation . . . under which were growing tons of *Begonia floribunda*, gold and silver colored, and many plants." He offered this description in a letter to his parents that was dated March 6, 1892. An enlargement of the male flower (♂) is shown (*opposite, upper left*), along with anthers (*bottom right*). The female (♀) flowers, with their distinctive coiled stigmas, are illustrated from various views.

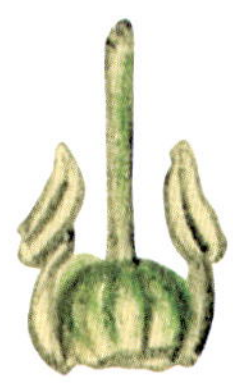

10.
Pyrola elliptica Nutt.
Waxflower Shinleaf
Ericaceae
Model 618 (1896)
R. Blaschka
Drawing no. 35
Colored pencil on paper
21 cm x 26.7 cm

In June 1895, Rudolf Blaschka collected this native evergreen, a member of the wintergreen family, in Cambridge, Massachusetts. The name, a diminutive of *Pyrus,* acknowledges the resemblance of the leaves to pear foliage that he characterized as "darkened leaf colors 20 and 21." Blaschka noted that the flowers are "white with greenish yellow shadowing slits of white around No. 4." He said that they usually have a five-chambered ovary, but he included a larger cross section with six chambers. The stem, he said, is "shiny and has stripes."

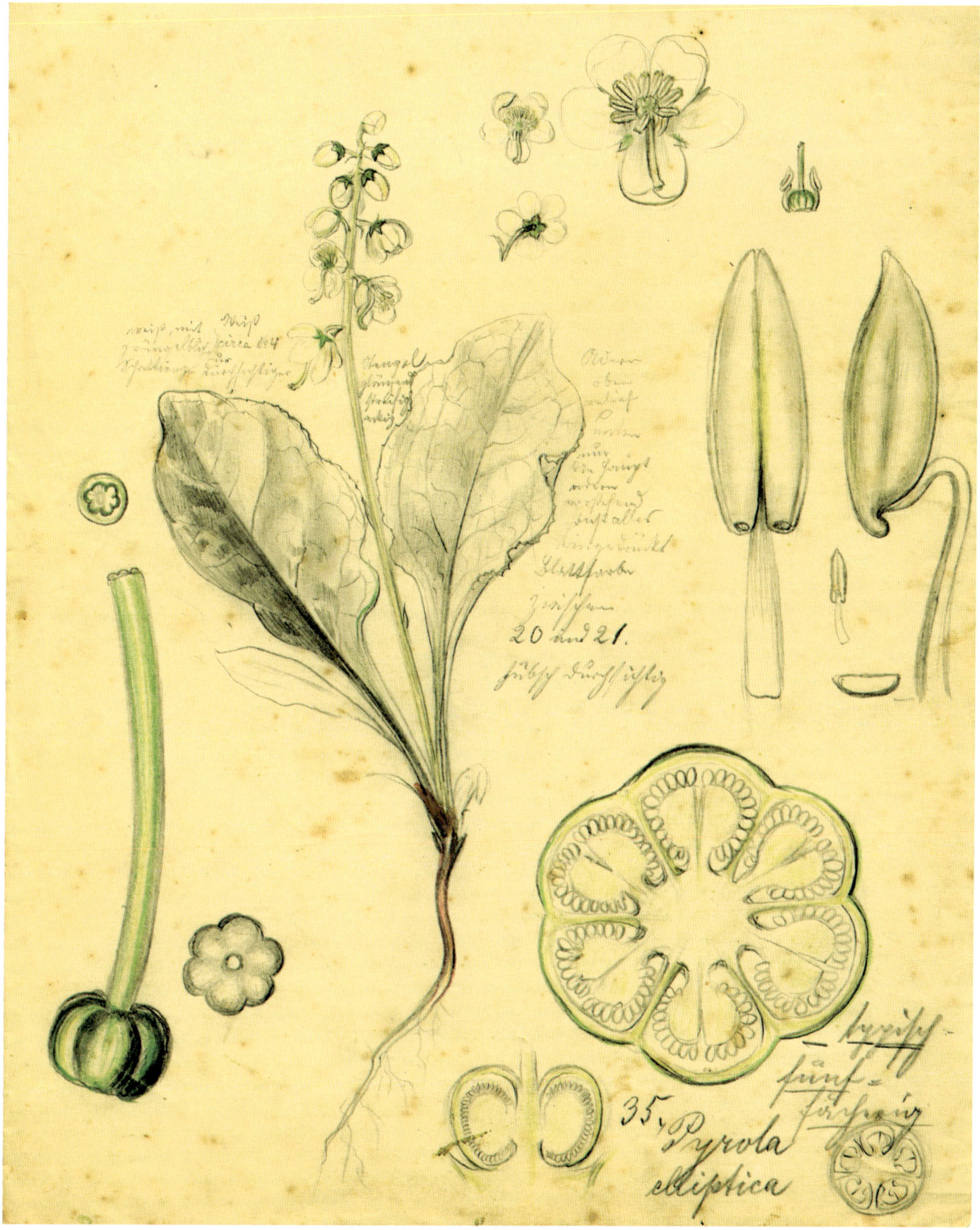
20 und 21.
35, Pyrola elliptica

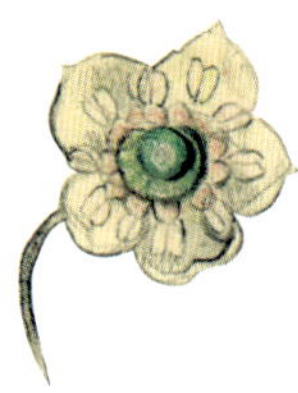

11.
Chimaphila umbellata (L.)
W. Bart.
Pipsissewa
Ericaceae
Model 629 (1898)
R. Blaschka
Drawing no. 36
Colored pencil on paper
21 cm x 26.7 cm

Like *Pyrola* in the drawing on page 119, this genus has violet stamens with "white hairs" (*upper right*). Rudolf Blaschka collected this evergreen perennial in Concord, Massachusetts, on June 30, 1895. The name of this low-growing, somewhat woody plant with dark leaves comes from the Greek for "winter" (*cheima*) and "to love" (*philein*). The drawing notes that "the colors of the leaves are 21–23a, strong and shiny on top, and underneath light, also shiny."

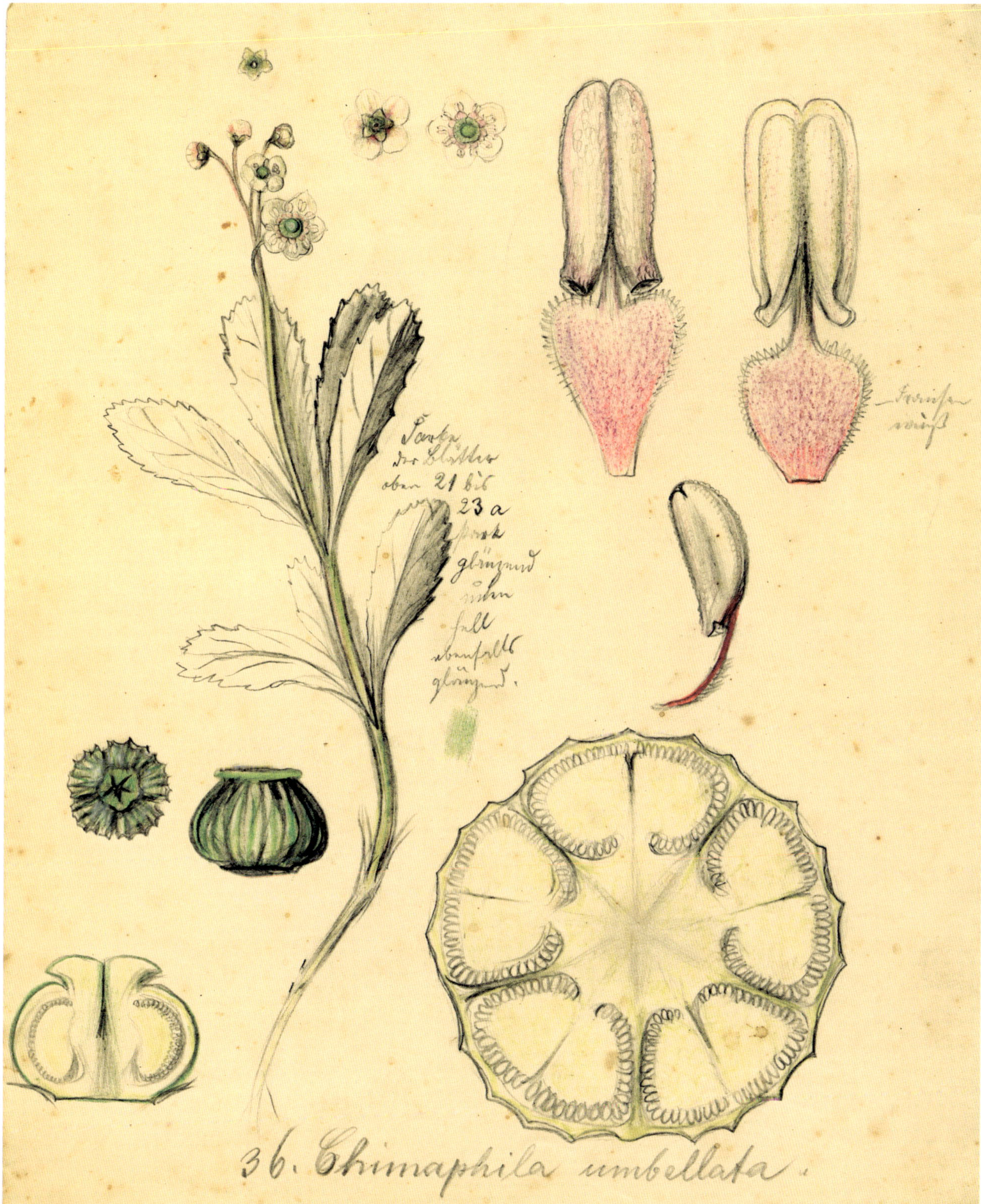
36. Chimaphila umbellata.

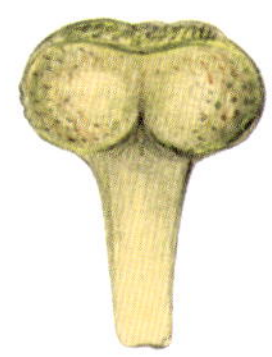

12.
Utricularia purpurea Walt.
Purple Bladderwort
Lentibulariaceae
Model 638 (1898)
R. Blaschka
Drawing no. 37
Colored pencil on paper
21 cm x 26.7 cm

On July 1, 1895, Rudolf Blaschka wrote to his parents about the carnivorous aquatics found in Concord, Massachusetts. He mentioned "2 species of *Utricularia*, a very strange water plant that is very difficult to force into bloom." The long, free-floating stems are described as "transparent, somewhat glasslike, and very finely haired," with bladder traps (*lower left*), which are also shown at the tips of some leaf segments. The delicate white flowers are "detailed very fine as the orchids."

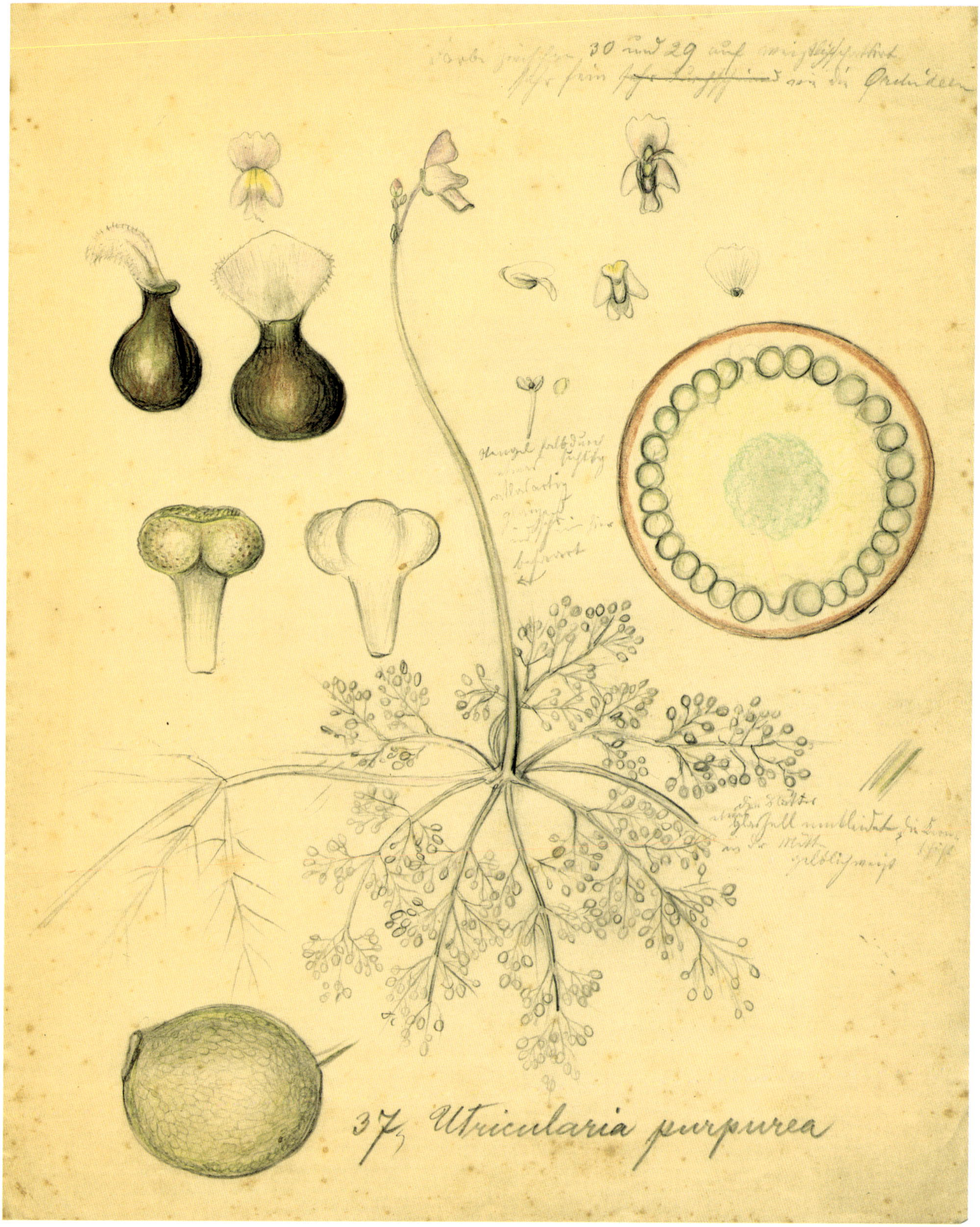
37, Utricularia purpurea

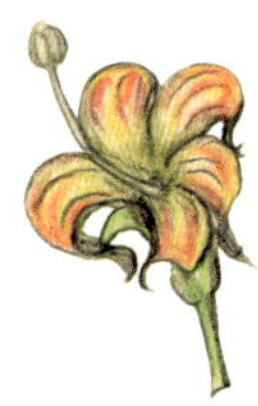

13.
Anacardium occidentale L.
Cashew
Anacardiaceae
Model 424 (1894)
L. and R. Blaschka
Drawing no. 44
Colored pencil on paper
21 cm x 26.7 cm

Native to tropical America, this leathery-leaved tree is widely cultivated for both the cashew apple and the cashew nut, which are shown in the lower drawings. The delicious pear-shaped "fruit" develops from the peduncle (stem) at the floral base. The true fruit, a smaller kidney-shaped structure, encases the seed (nut) that grows below. In Jamaica, Rudolf Blaschka illustrated the flowers, the fruit, and the foliage that he described as "half matte yellow green," a color he compared with that of "walnut leaves."

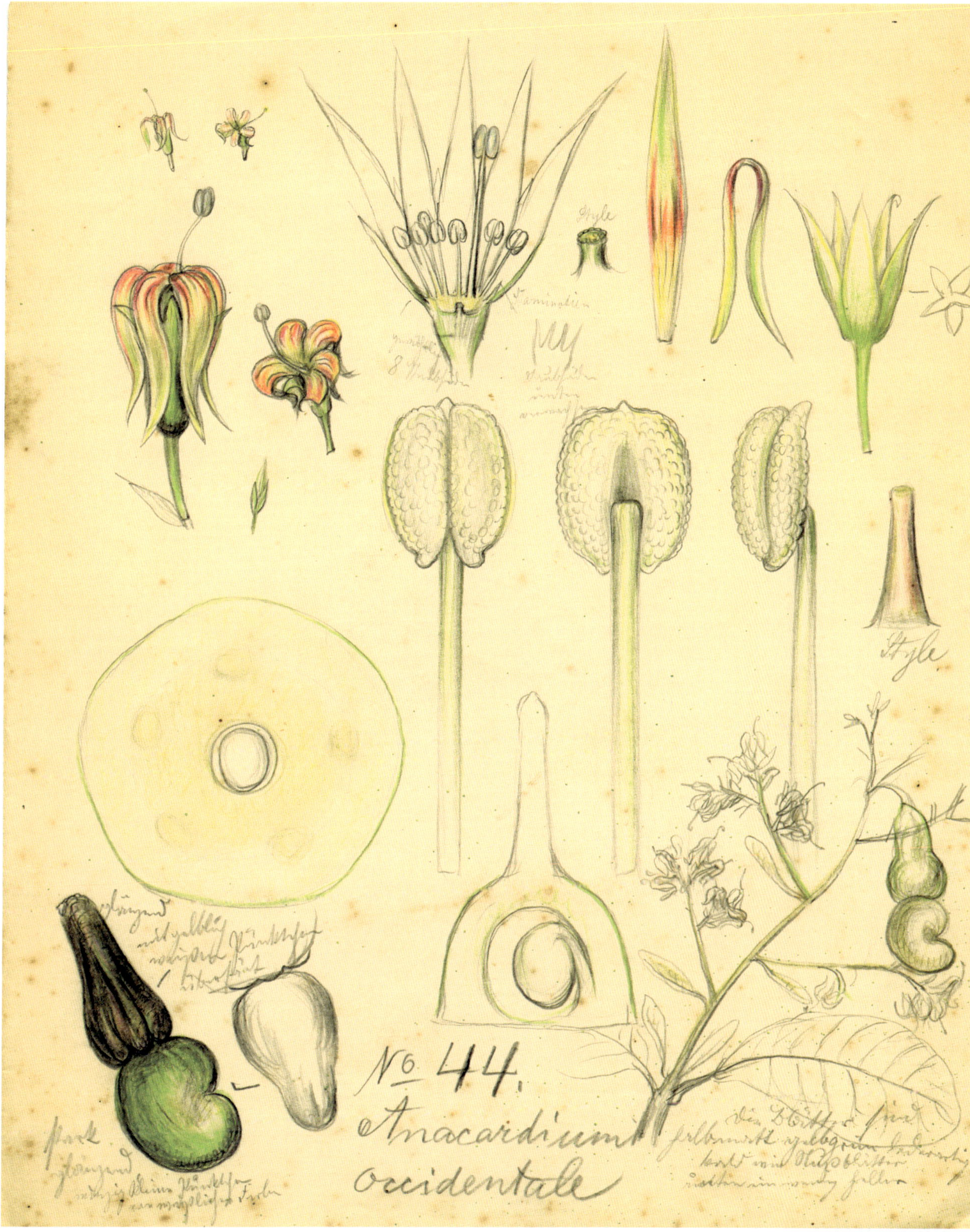
Style
Style
No 44.
Anacardium occidentale

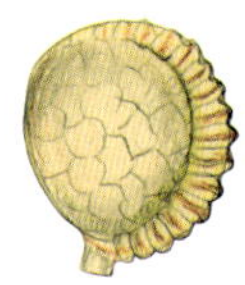

14.
Gymnogramme rufa L.
[currently *Hemionitis rufa* (L.) Sw.]
Fern
Pteridaceae
Model 709 (1903)
R. Blaschka
Drawing no. 48
Colored pencil on paper
21 cm x 26.7 cm

Not all of the models in the Ware Collection are of flowering plants. This beautiful fern, which Rudolf Blaschka described as "completely occupied with fine hairs" that are "wooly brown," shows the whole frond (*left*). The enlarged underside of a pinna illustrates the spore-bearing structures (sporangia) along the veins. The enlarged side view of a single sporangium shows the annulus (*right*), a ribbon of cells on the right, that will contract and release spores. Blaschka collected the fern in Jamaica in 1892.

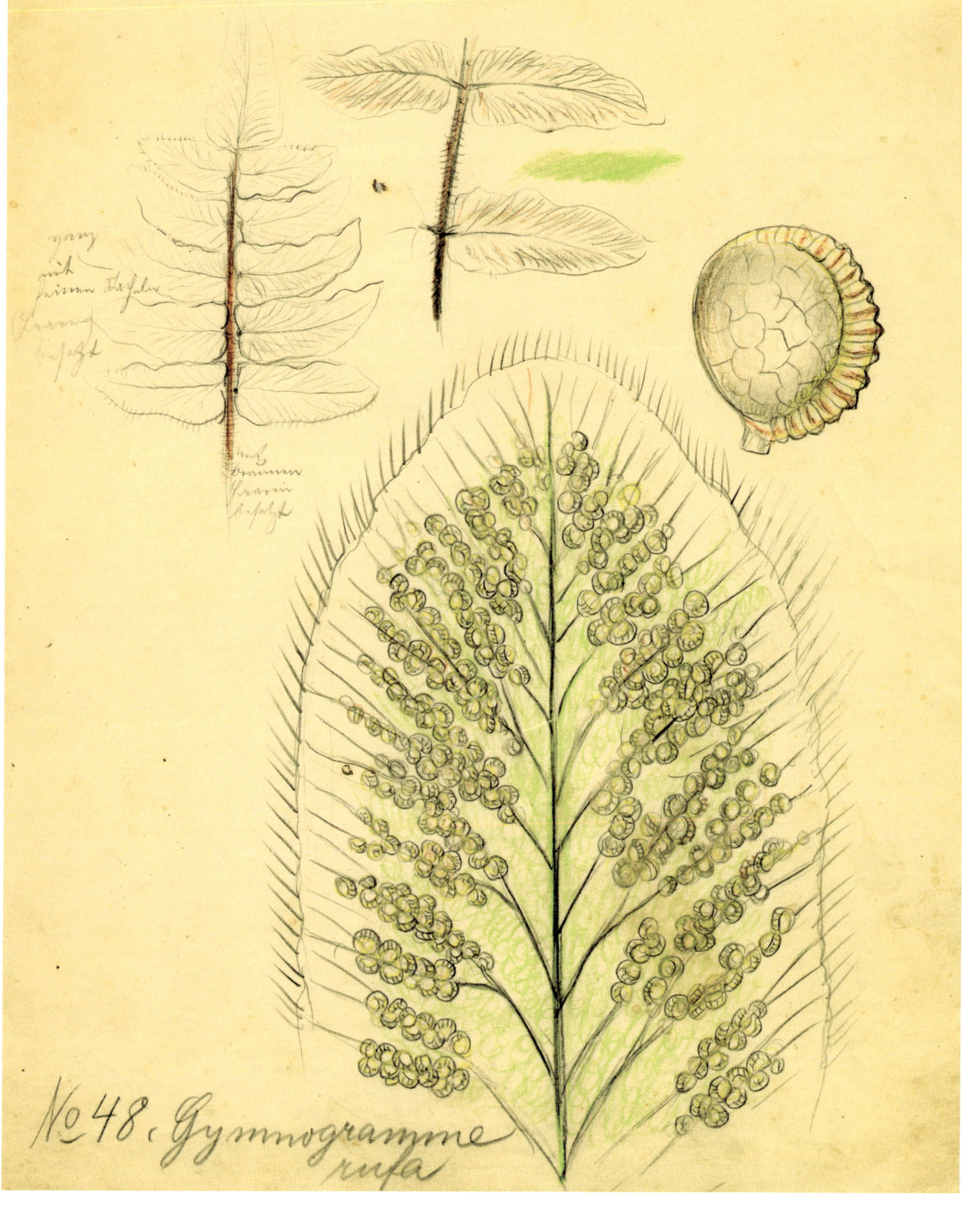
No 48. Gymnogramme rufa

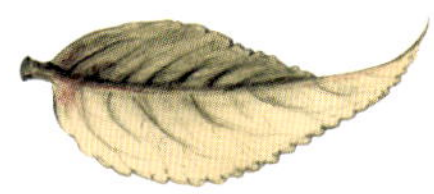

15.
Monarda didyma L.
Scarlet Bee-balm
Lamiaceae
Model 698 (1903)
R. Blaschka
Drawing no. 51
Colored pencil on paper
21 cm x 26.7 cm

This perennial native herb, which Rudolf Blaschka collected in Southwick, Massachusetts, on July 4, 1895, has become a beloved garden plant. He described the color of the bright red flowers as "carmine red lacquer, half-matte," and he provided a color swatch (*right*). The whorl of flowers, typical of the mint family, develops around a square stem that is "half shiny, color no. 20" (green) with "purple edges." The enlarged anthers and pistil are shown below the dissected flower.

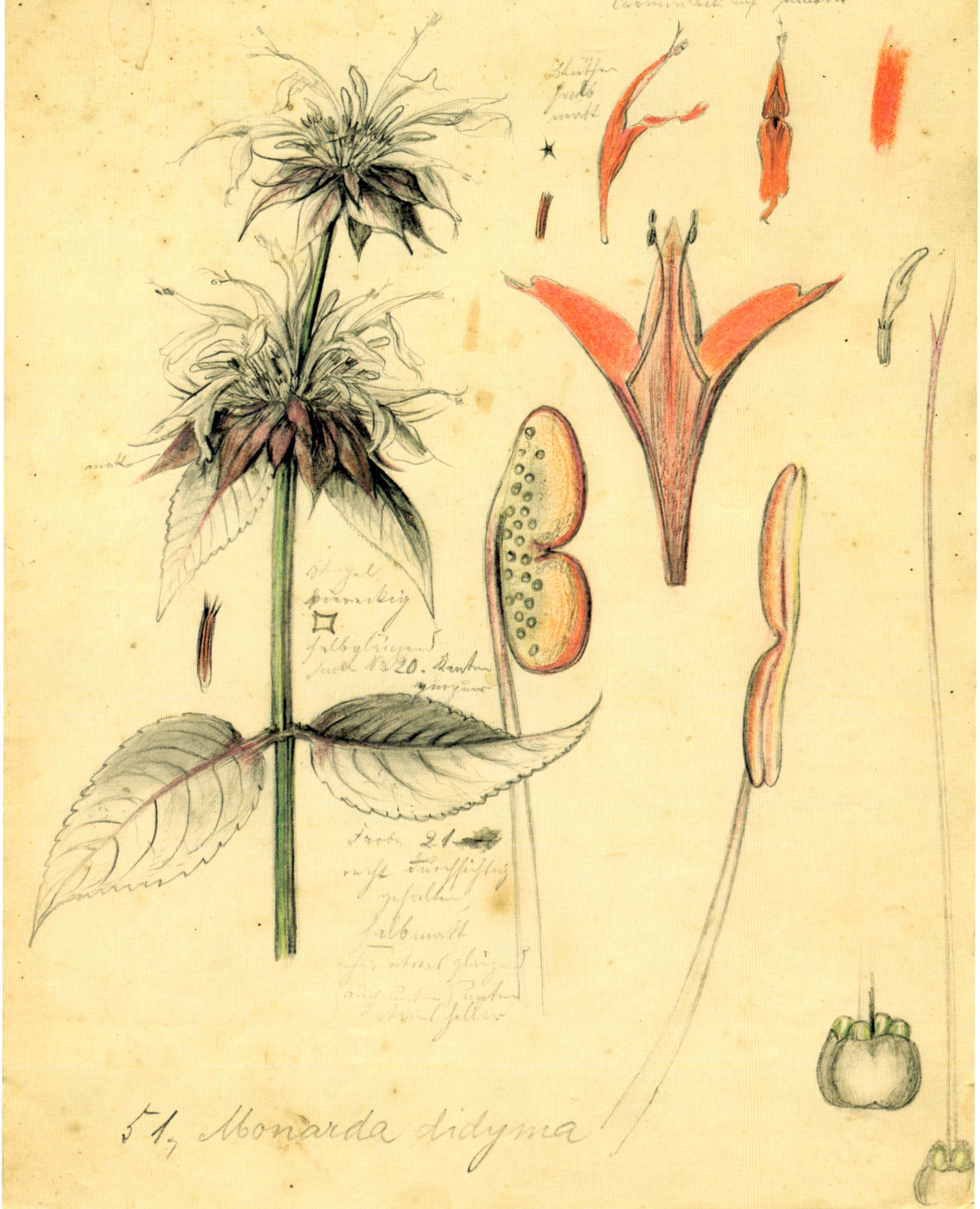
51, Monarda didyma

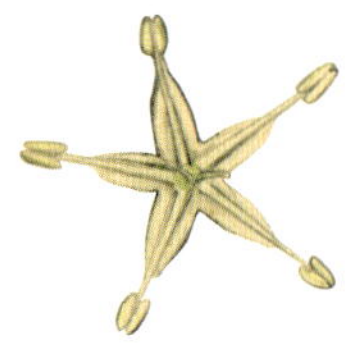

16.
Salsola kali L.
Saltwort
Chenopodiaceae
Model 650 (1899)
R. Blaschka
Drawing no. 61
Colored pencil on paper
21 cm x 26.7 cm

Rudolf Blaschka collected this introduced annual in Nahant, Massachusetts, on July 7, 1895. The plant ash, an important source of alkali salts, was used as a flux in glassmaking to reduce the melting point of the silica. The branching hairy stems have stiff, prickly leaves. Minuscule flowers, without petals, have a five-part calyx and aligned stamens that develop above two small bracts in the leaf axil, as shown in the upper drawings.

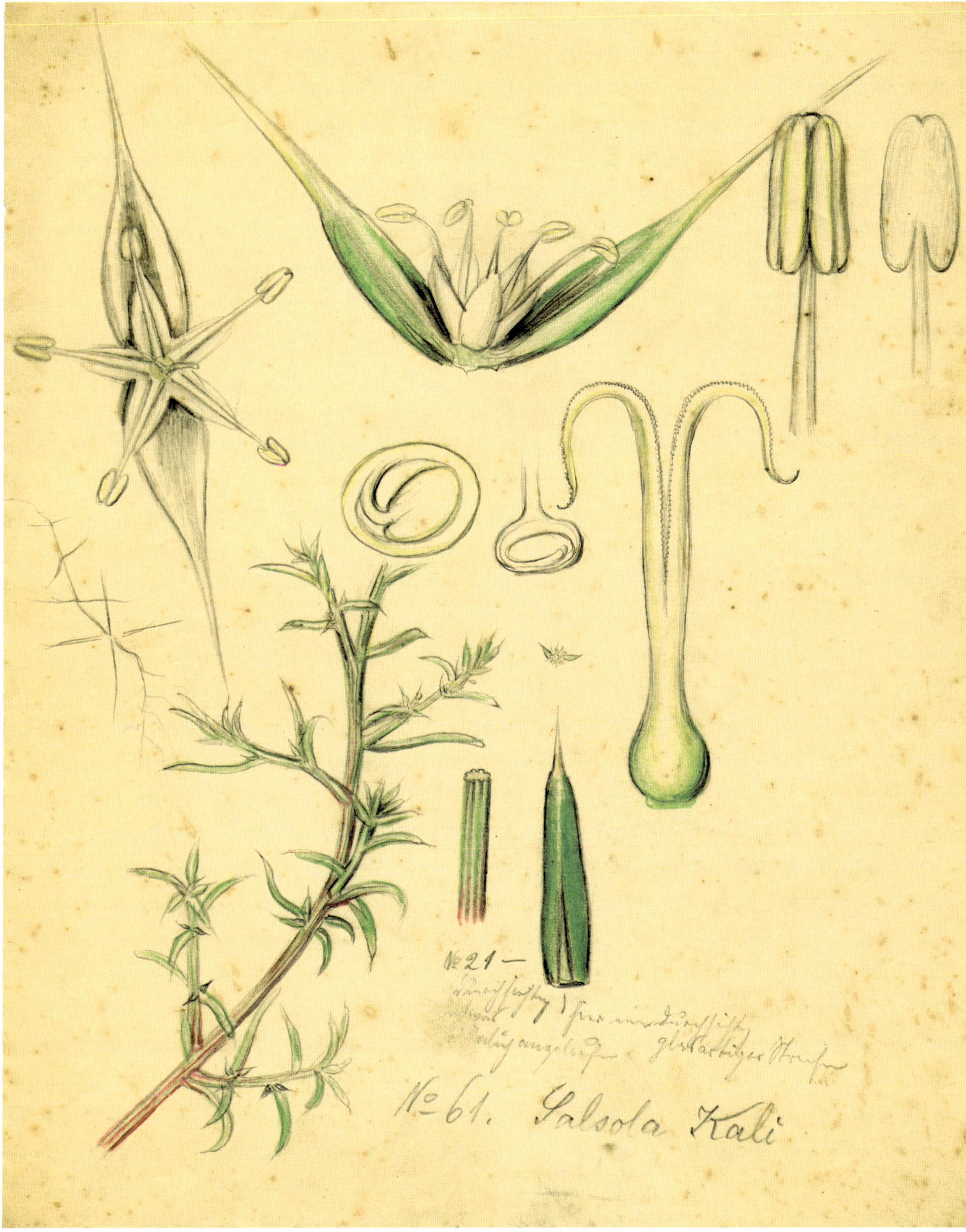
№ 21 —
№ 61. Salsola Kali

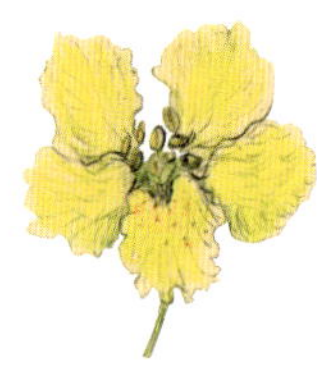

17.
Parkinsonia aculeata L.
Jerusalem Thorn
Fabaceae
Model 428 (1894)
L. and R. Blaschka
Drawing no. 62
Colored pencil on paper
21 cm x 26.7 cm

This tree is native to tropical America and widely cultivated in the South. Rudolf Blaschka collected specimens in Jamaica and recorded his detailed studies. He described the flowers (*upper left*) as "very delicate yellow, transparent" and "wrinkled along their length." The petals (*middle right*) are "hairy on the inside." The texture of the stem (*lower left*) is "as if dusty," and the color, along with that of the leaves, is described as "gray-green matte."

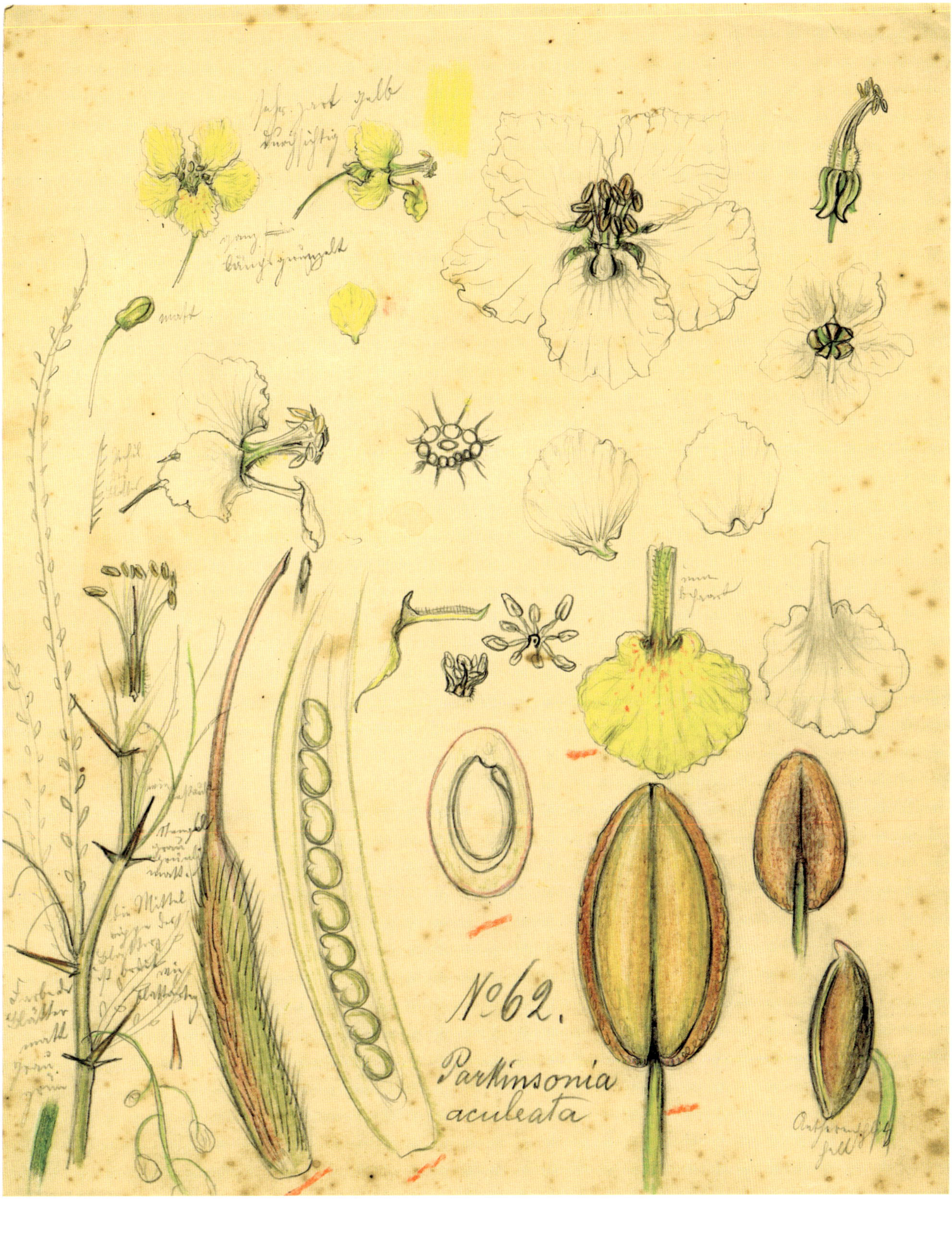
No 62.
Parkinsonia
aculeata

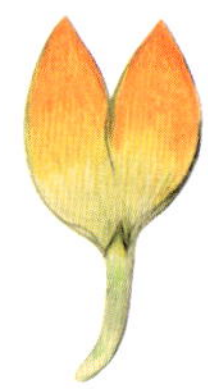

18.
Polygala lutea L.
Orange Milkwort
Polygalaceae
Model 660 (1899)
R. Blaschka
Drawing no. 62
Colored pencil on paper
21 cm x 26.7 cm

Rudolf Blaschka drew this native biennial in Suffolk, Virginia, on July 13, 1895. The individual flowers in the inflorescence, he said, are "very loud orange toward the tip." The winged petals are shown in several views (*middle*), with the various parts enlarged (*right*). The pistil (*center*) has a tufted stigma and an anther attached underneath, which is also shown enlarged (*right*).

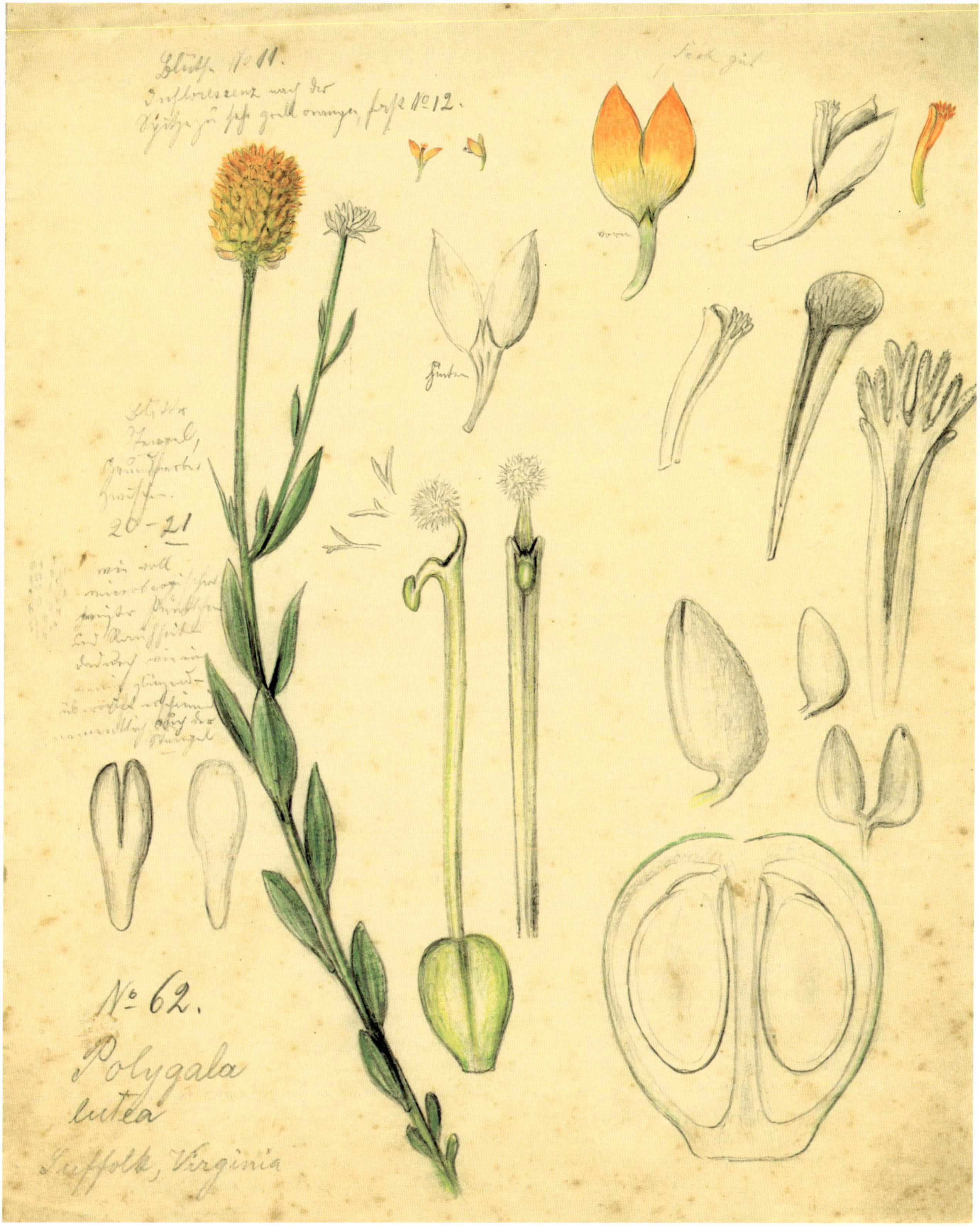
20–21
No 62.
Polygala
lutea
Suffolk, Virginia

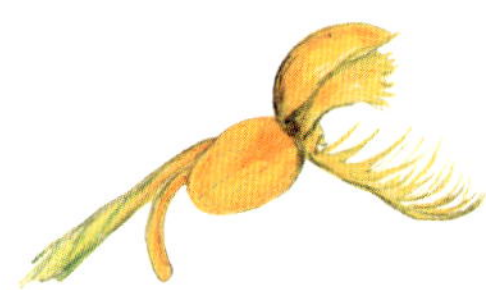

19.
Habenaria ciliaris (L.)
R. Br. ex Ait. f.
[currently *Platanthera*
ciliaris (L.) Lindl.]
Yellow Fringed Orchid
Orchidaceae
Model 642 (1898)
R. Blaschka
Drawing no. 66
Colored pencil on paper
21 cm x 26.7 cm

On July 15, 1895, Rudolf Blaschka collected this spurred orchid in Virginia. It features a fringed lower petal, which is shown in various views. The drawings to the right of the leaves show the interior structure of the flower. A pair of pollinaria, each of which resembles a string of balloons, contain individual pollen grains in what Blaschka called a "juice like yellowish water glass." The sticky end of the stipe becomes attached to an insect, which then distributes pollen to other flowers.

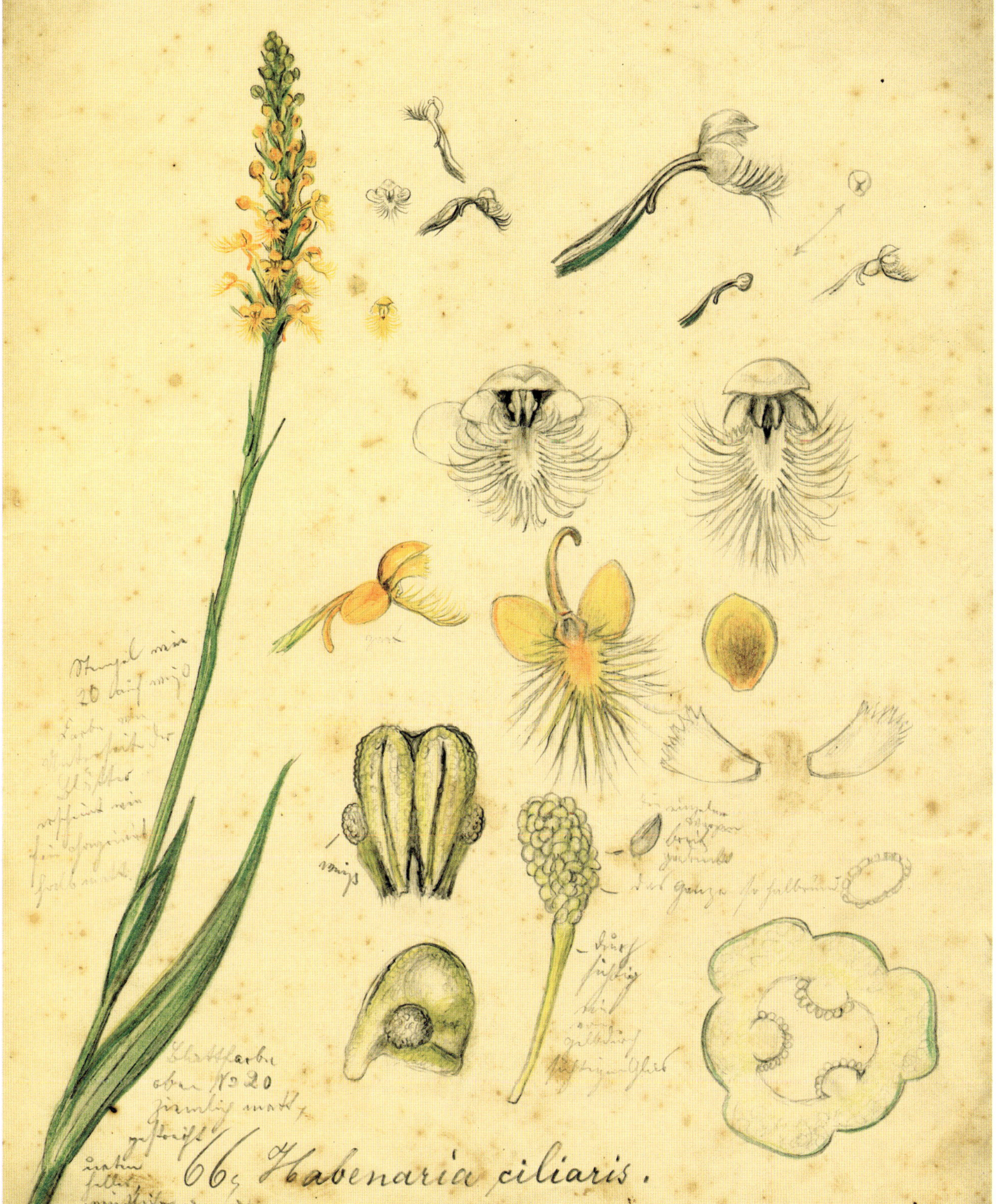
66, Habenaria ciliaris.

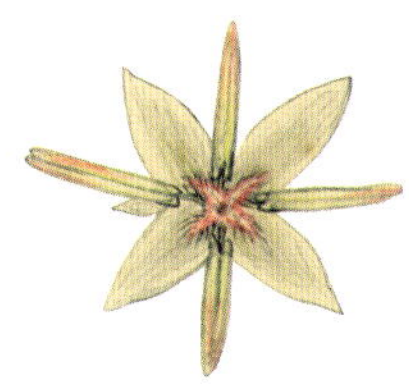

20.
Myriophyllum heterophyllum Michx.
Water Milfoil
Haloragaceae
Model 641 (1898)
R. Blaschka
Drawing no. 71
Colored pencil on paper
21 cm x 26.7 cm

This native aquatic, collected in Suffolk, Virginia, on July 17, 1895, has a broad distribution. The plant produces three types of flowers along its stout stem, male (♂), female (♀), and those with both sex organs. The purple pistils are shown (*middle*), with a cross section of the ovary below them. The colorful anthers are to the left of the plant. The delicate minuscule leaves, in groups of four or six, are described as "somewhat shiny on top; underneath not lighter."

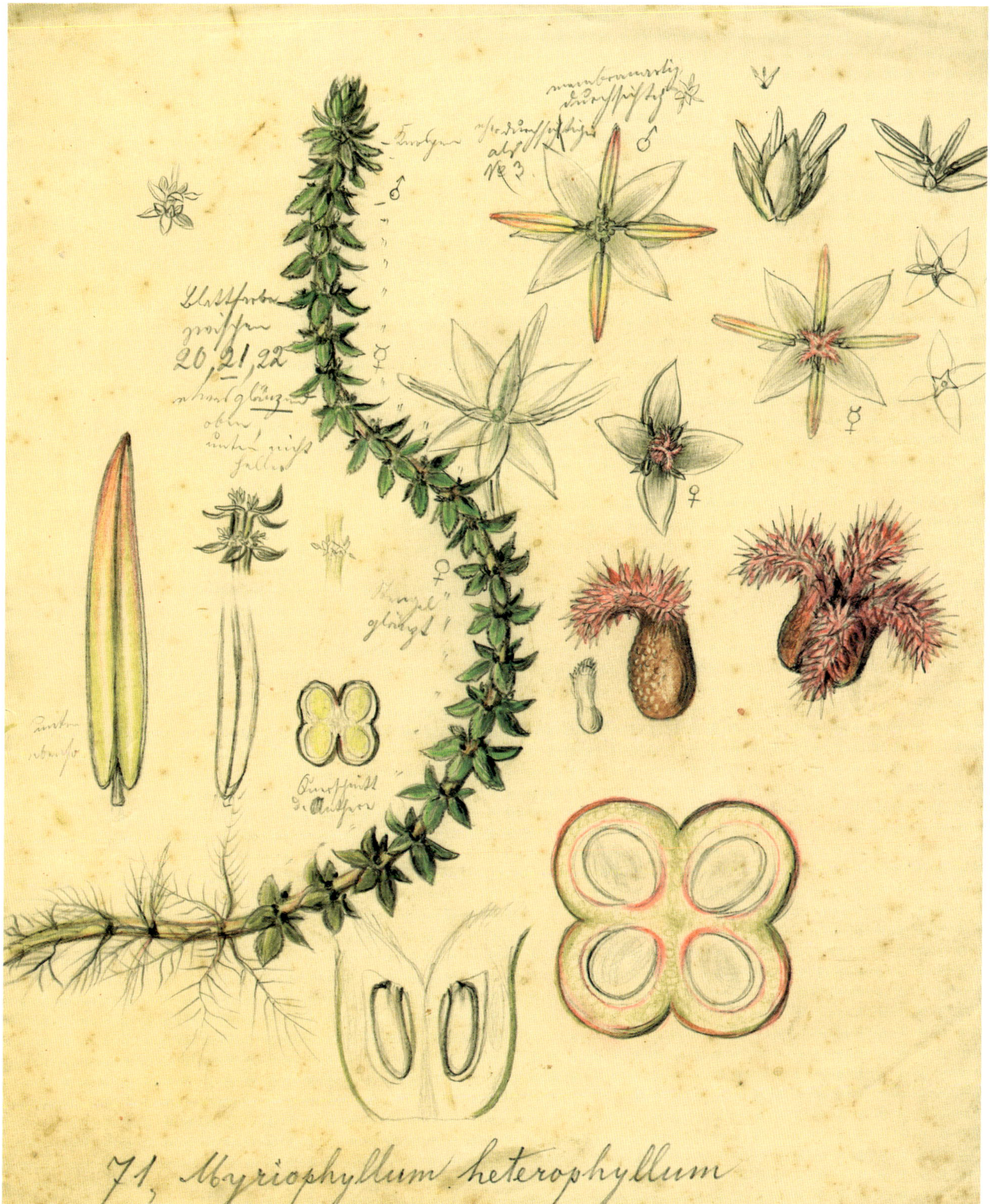
71, Myriophyllum heterophyllum

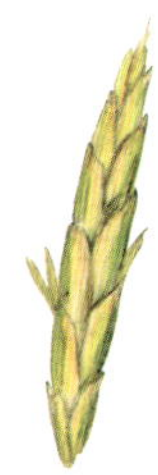

21.
Cyperus strigosus L.
Straw-Colored Flat Sedge
Cyperaceae
Model 654 (1899)
R. Blaschka
Drawing no. 72
Colored pencil on paper
21 cm x 26.7 cm

Although similar to grass, a sedge is distinguished by its triangular stem (*shown in cross sections to the left of the plant*) and differences in its flowers and fruits. An individual inflorescence (*left*) shows the middle flowers with extended anthers that release pollen into the wind, to be deposited on the three-part stigma of the pistil (*right*). Rudolf Blaschka collected this specimen, which is native to the New World, in Virginia in 1895.

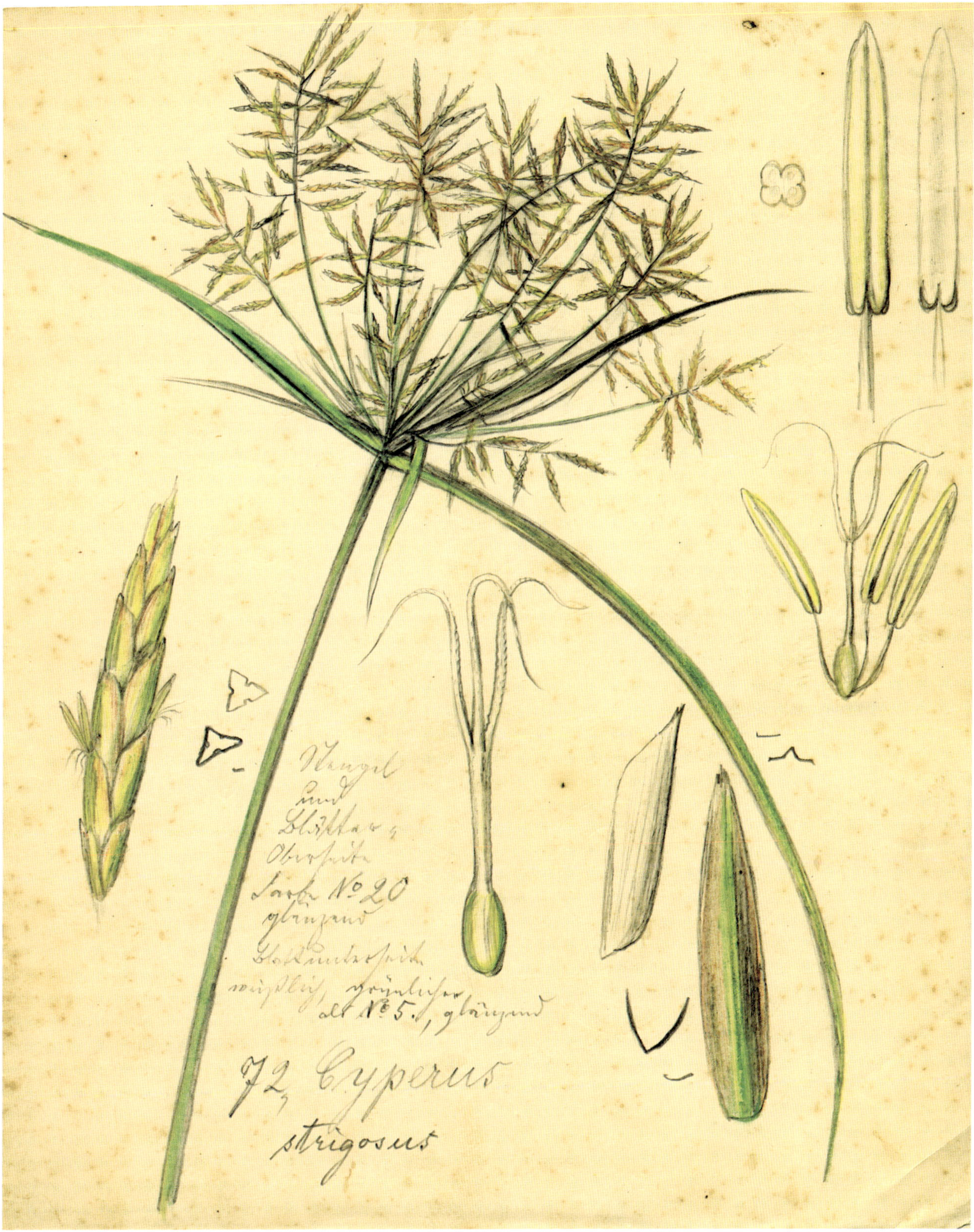
72, Cyperus
strigosus

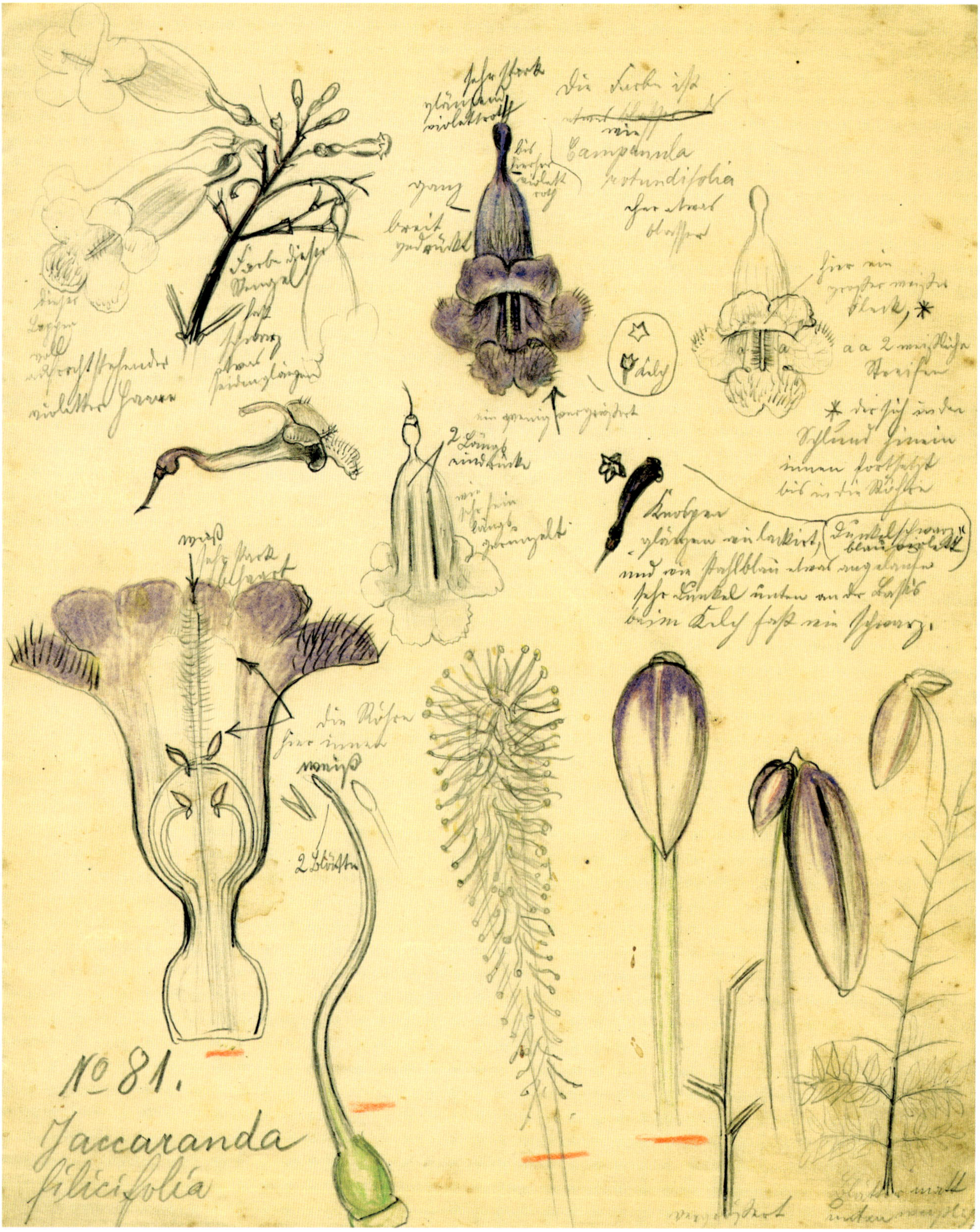
Campanula
rotundifolia
No 81.
Jaccaranda
filicifolia

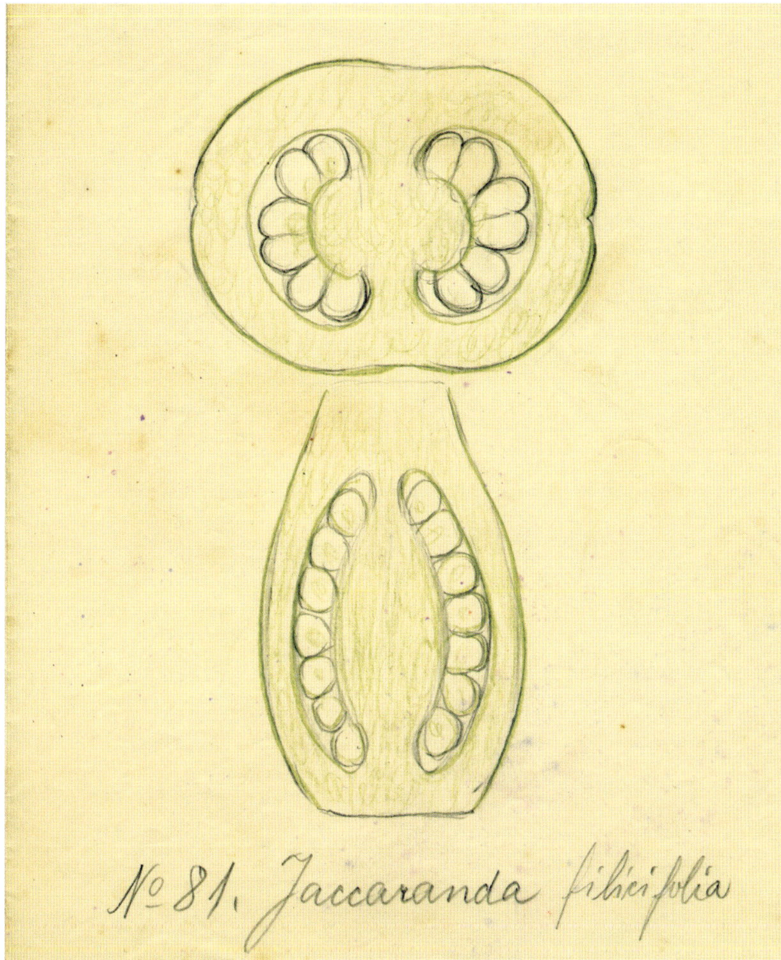

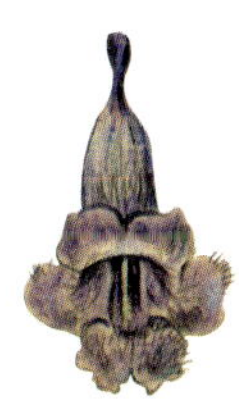

22.
Jaccaranda filicifolia
(*Jacaranda filicifolia* D. Don)
Fern Tree
Bignoniaceae
Model 550 (1895)
L. and R. Blaschka
Drawing no. 81
Colored pencil on paper
21 cm x 26.7 cm

Rudolf Blaschka documented this tropical species from Castleton Gardens, Jamaica, in 1892. He described it as having "very strongly shiny red-violet" tubular flowers with "lips full [of] vertically standing violet hair" and black stems "somewhat shiny like silk." The curved anthers, sketched as close-ups (*opposite, bottom right*), are placed in the enlarged flattened flower. The pistil has been removed and placed to the right; the showy "tube, white inside" is a sterile filament.

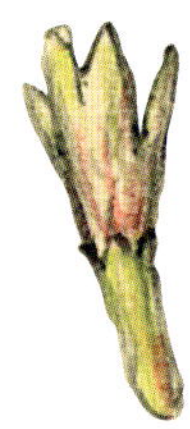

23.
Pithecolobium saman
Jacquin [currently *Samanea saman* (Jacq.) Merr.]
Rain Tree
Fabaceae
Model 442 (1894)
L. and R. Blaschka
Drawing no. 96
Colored pencil on paper
21 cm x 26.7 cm

Native to tropical America, this ornamental shade tree is cultivated worldwide. In Jamaica, Rudolf Blaschka carefully dissected the central showy flower, which he characterized as "usually extra large and fertile." The frilly filaments of the anthers, "about 30, grown into [a] tube" at the base of the flower, are sketched at the bottom. Blaschka provided details about the "finely haired" branch sculpted into "5 furrows." The "leaf stem, thickened at the base, flows into 3 furrows."

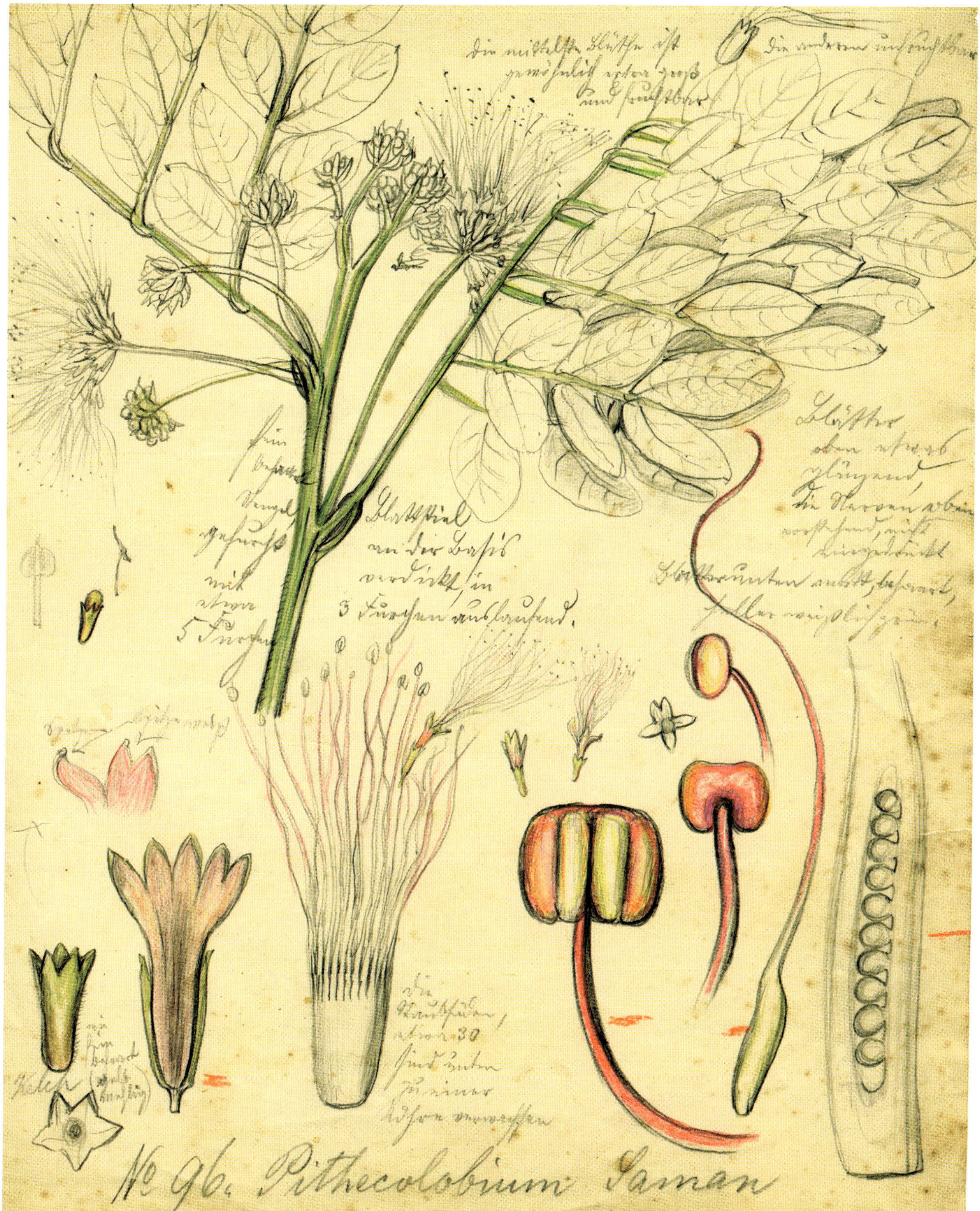
Kelch
No 96. Pithecolobium Saman

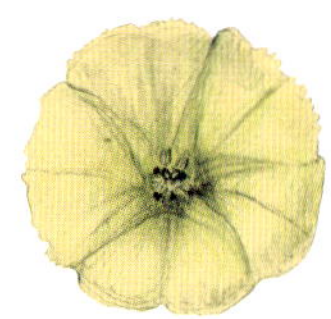

24.

Convolvulus occidentalis A. Gray
[currently *Calystegia occidentalis* (A. Gray) Brummitt]
Western Bindweed
Convolvulaceae
Model 505 (1894)

L. and R. Blaschka
Drawing no. 155
Colored pencil on paper
21 cm x 26.7 cm

This native perennial vine survives on the dry hills of Oregon and California. It was collected in San Diego. Rudolf Blaschka compared the matte color of the leaves to "our Winden [morning glories]." The stem is "grooved and twisted." The stripes on the flower's exterior are "strongly shiny, as if lacquered," whereas the rest is "somewhat more matte and whiter." The "delicate yellowish white" interior is "somewhat greenish, effortlessly shiny like royal silk."

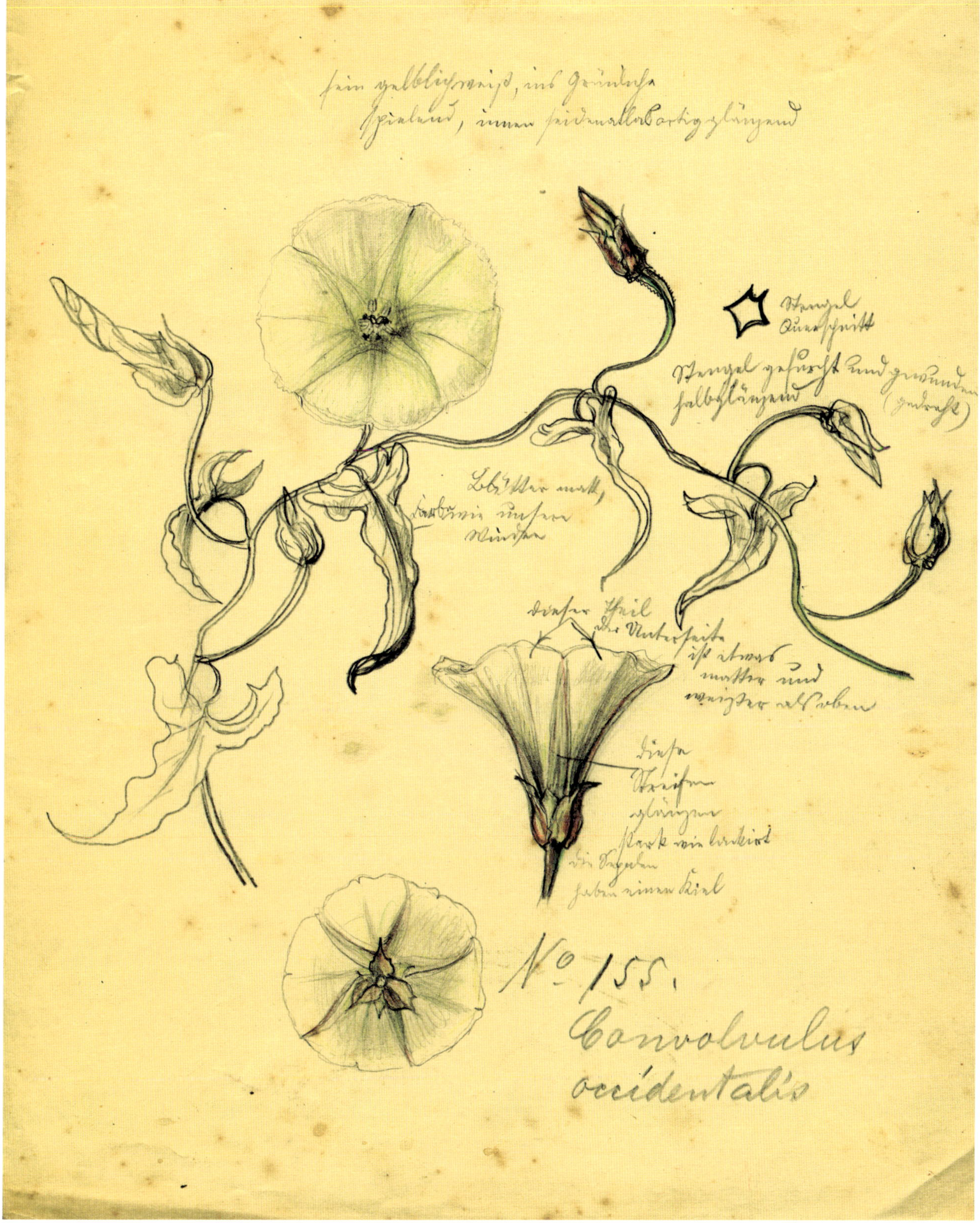
No. 155.
Convolvulus
occidentalis

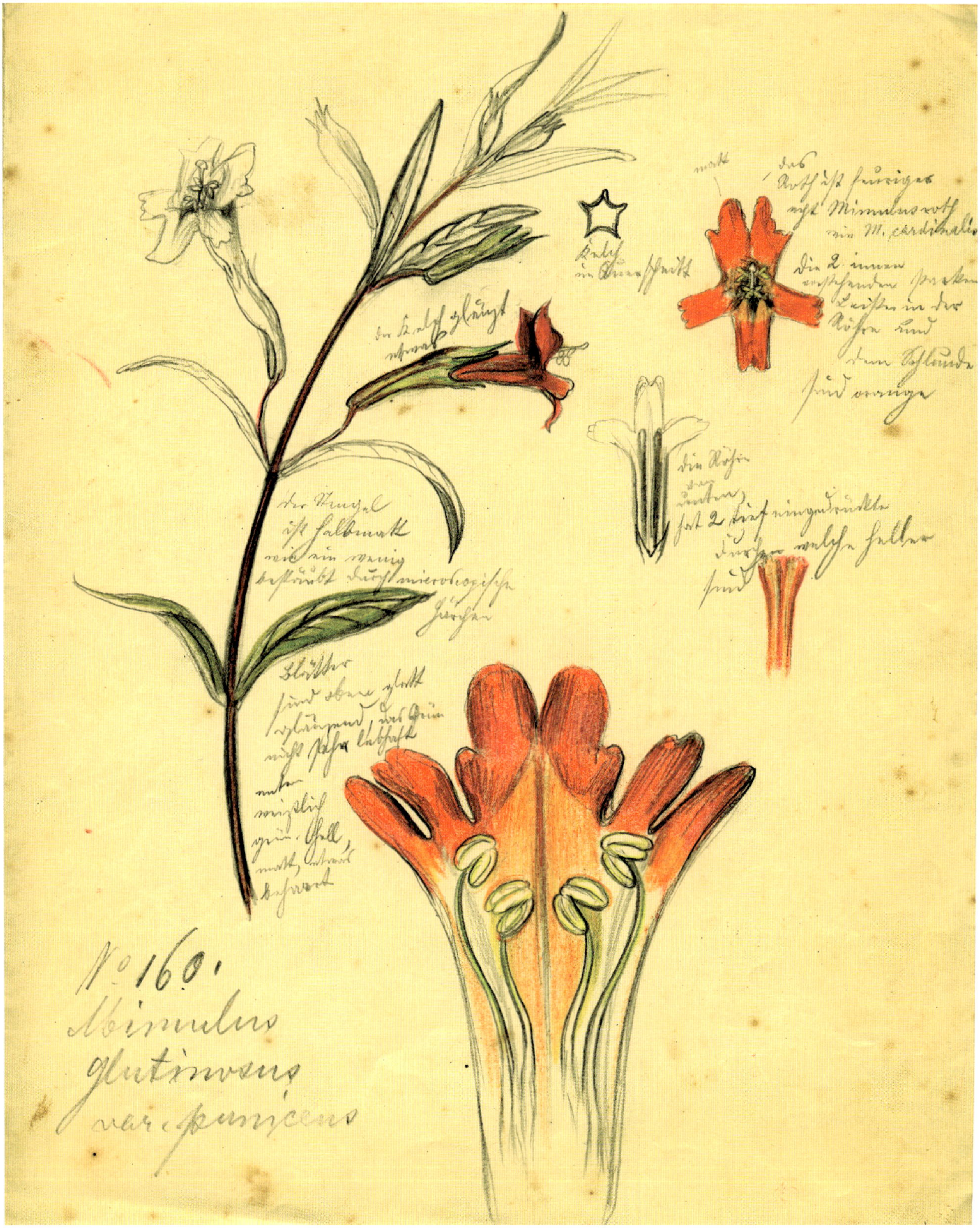
No 160.
Mimulus
glutinosus
var. puniceus

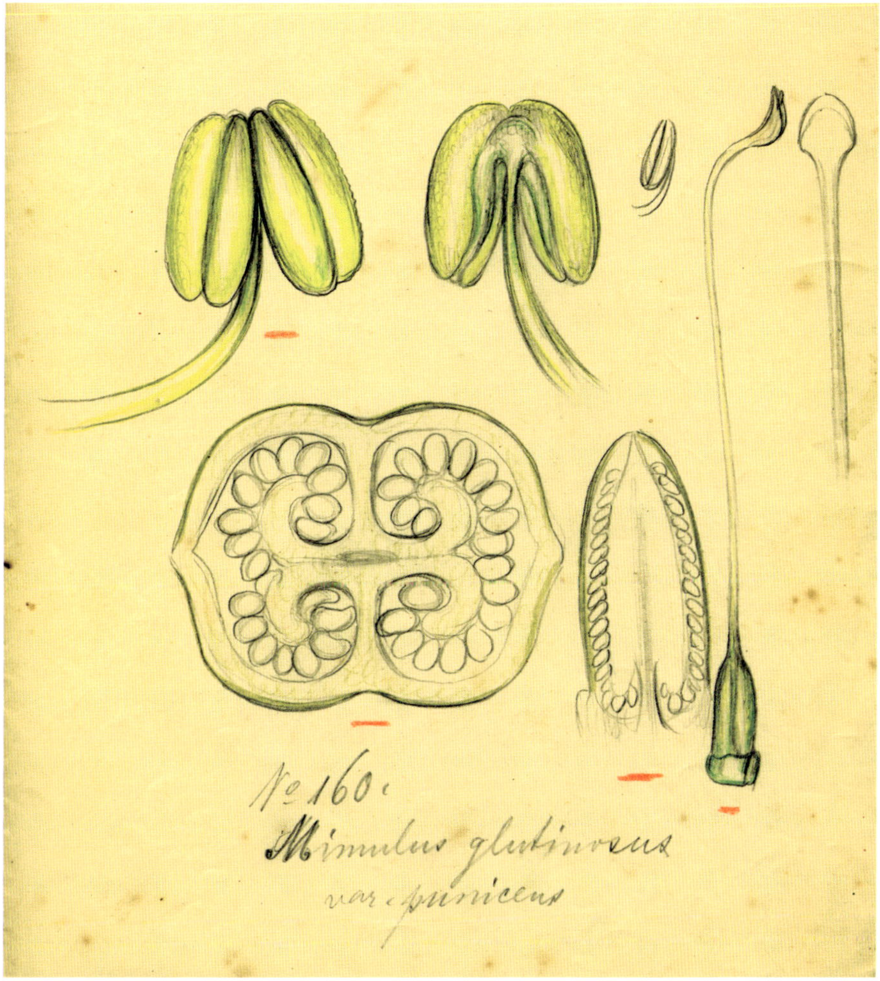

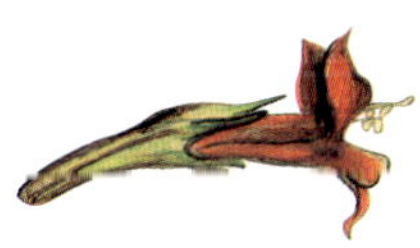

25.
Mimulus glutinosus
J. C. Wendl. var. *puniceus*
(Nutt.) A. Gray (currently
Diplacus puniceus Nutt.)
Red Bush Monkey Flower
Scrophulariaceae
Model 497 (1894)
L. and R. Blaschka
Drawing no. 160
Colored pencil on paper
21 cm x 26.7 cm

This "fiery" red flower has "two inner, strongly protruding dark ledges in the tube" that, along with the throat, are orange. The native shrub was collected in 1892 near San Diego, California, where Rudolf Blaschka made acute observations about the "half-matte" stem "dust covered because of microscopic hairs." The upper surface of the smooth, shiny leaves is a green that he described as "not very lifelike." Underneath, he said, the leaves are "whitish green, light, matte, somewhat hairy."

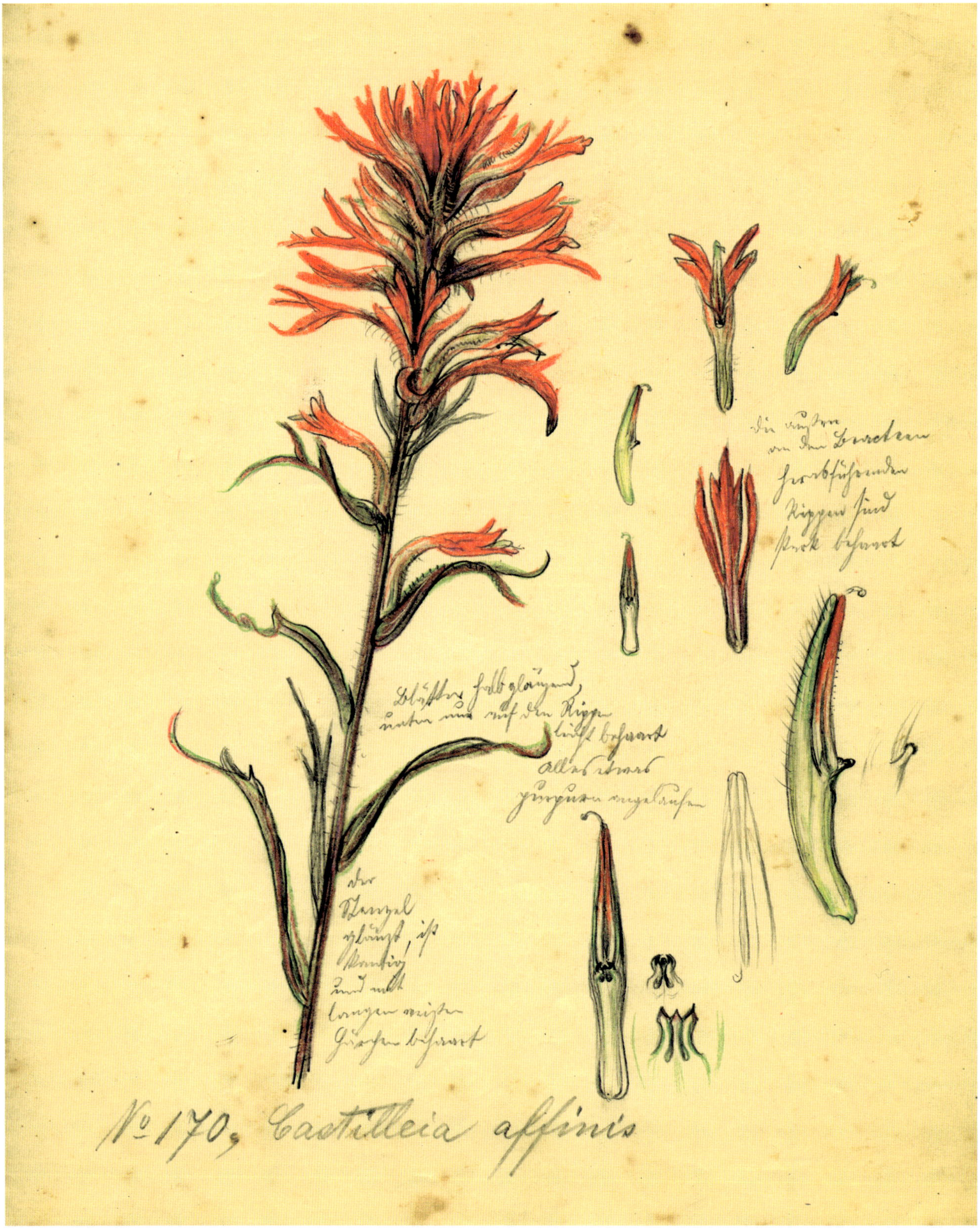
No 170, Castilleia affinis

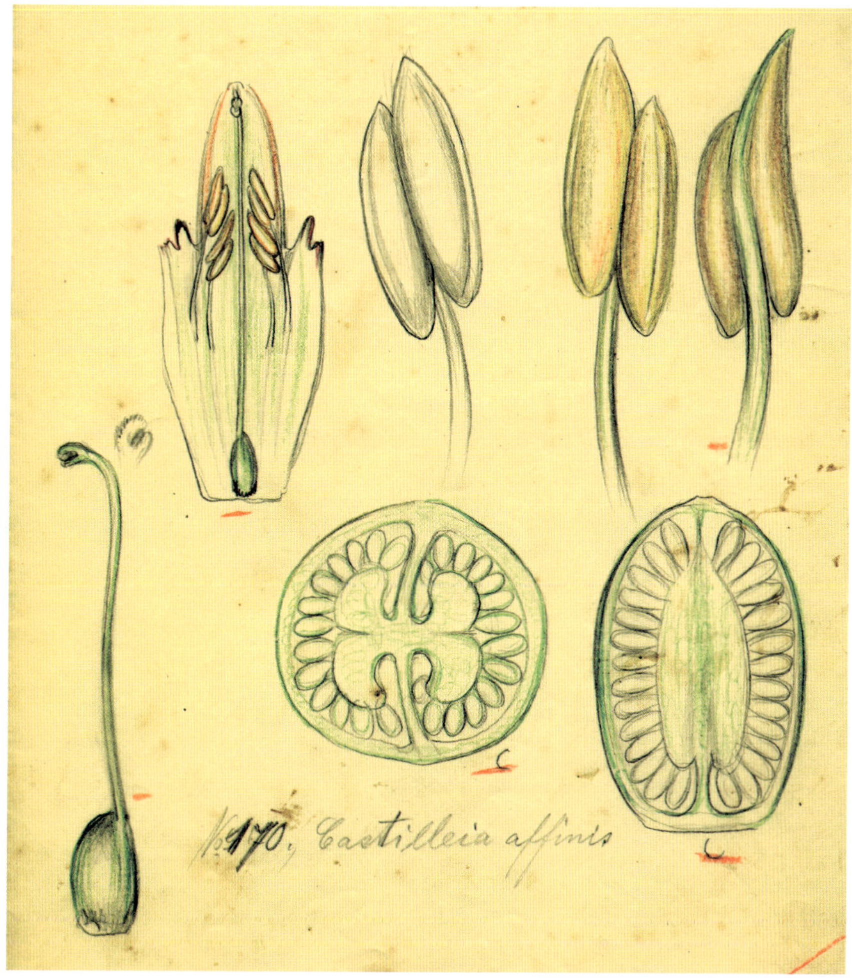

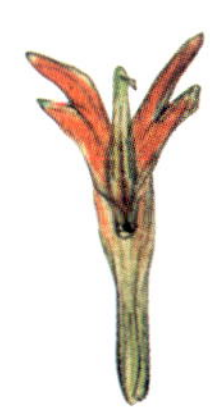

26.
Castilleia affinis (*Castilleja affinis* Hook. and Arn.)
Indian Paintbrush
Scrophulariaceae
Model 500 (1894)
L. and R. Blaschka
Drawing no. 170
Colored pencil on paper
21 cm x 26.7 cm

"It was a great pleasure to wander around the neighboring grounds and botanize, during which time I found some plants growing wild in great numbers that we at home raise in the garden so carefully. Some examples are . . . *Castilleia* species with their fiery red bracts." So wrote Rudolf Blaschka to his parents from San Diego, California, on April 30, 1892. "Long white hairs" cover the stem, he noted, and the leaves are "lightly hairy" to reduce water loss and "somewhat hazy purple."

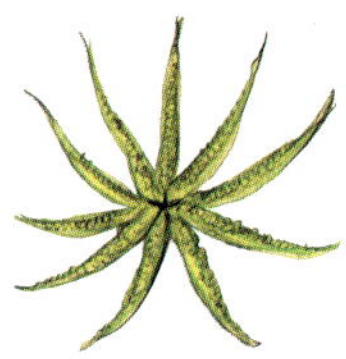

27.
Mesembryanthemum aequilaterale Willd.
[currently *Carpobrotus aequilateralis* (Willd.) S. T. Blake]
Sea Fig
Aizoaceae
Model 400 (1893)
L. and R. Blaschka
Drawing no. 185
Colored pencil on paper
21 cm x 26.7 cm

These fragrant flowers, described by Rudolf Blaschka as "delicately fuchsia-violet," develop into edible fruits, which are shown here in cross and longitudinal sections. In a letter to his parents that is dated April 25, 1892, he wrote of "wheat fields, vineyards, rose gardens, and everywhere the rich wild vegetation of blue lupines, delphinium, *Mesembryanthemum*, and a mass of yellow composites that often give entire hills a lively yellow color." He collected this specimen in California, where the introduced shrub grows along the seacoast and in salty soils.

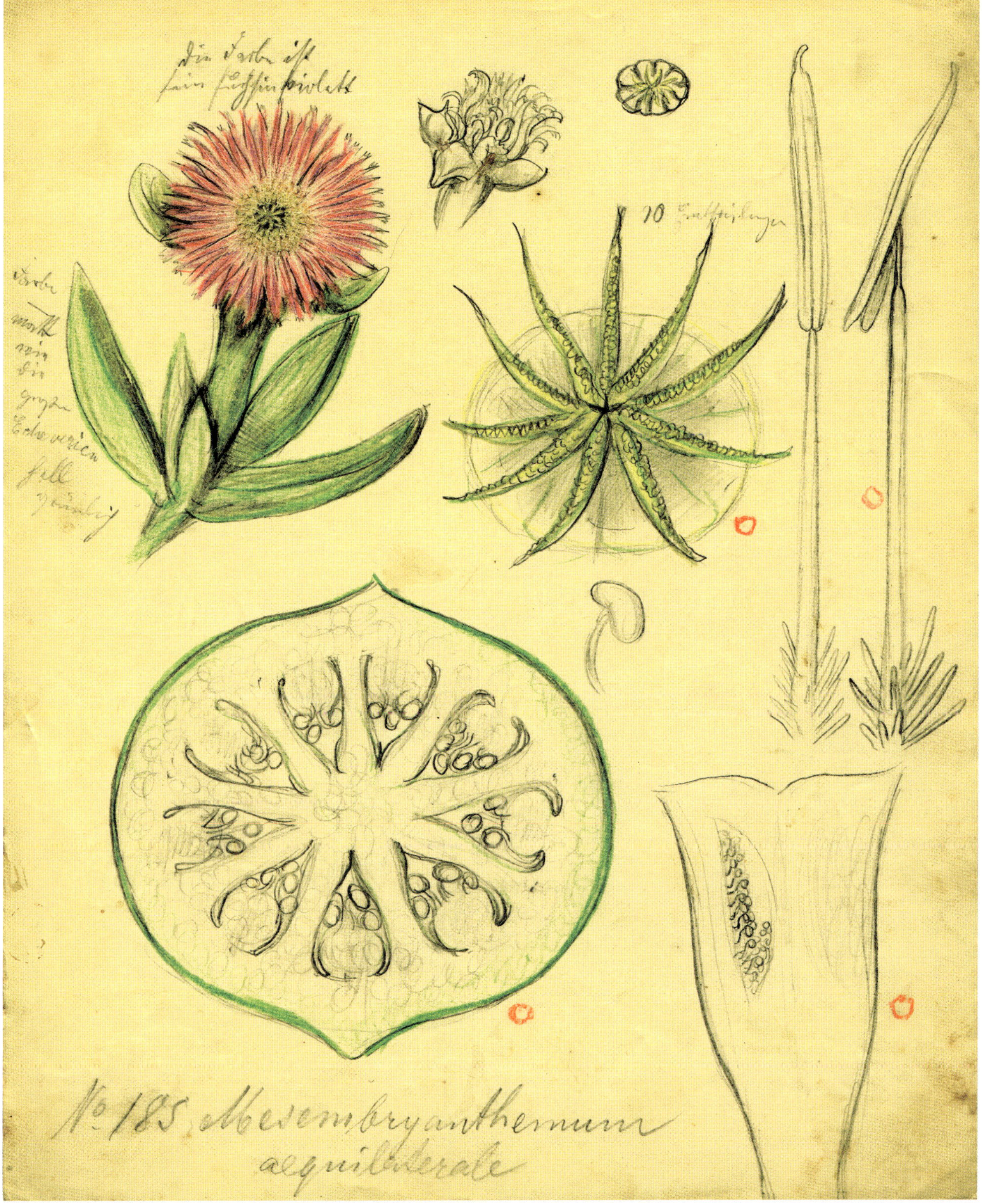
No 185 Mesembryanthemum
aequilaterale

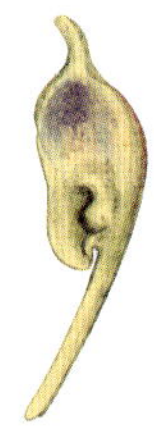

28.
Oxytropis lamberti
(*O. lambertii* Pursh)
Purple Locoweed
Fabaceae
Model 438 (1894)
L. and R. Blaschka
Drawing no. 243
Colored pencil on paper
21 cm x 26.7 cm

The pealike flower of this perennial, which is native from Saskatchewan to Texas, was collected in Colorado Springs, Colorado, in 1892. Rudolf Blaschka is poetic in his description: "The color is delicate white like white rose, with an unusually delicate greenish yellow tinge. The flower parts [petals] are very thin calcium red, greenish yellow [with] a reddish tinge. The flags are also somewhat longer, also very rosy rimmed; the calyx teeth are brown red." The entire plant is "thickly covered with long hair that appears silklike."

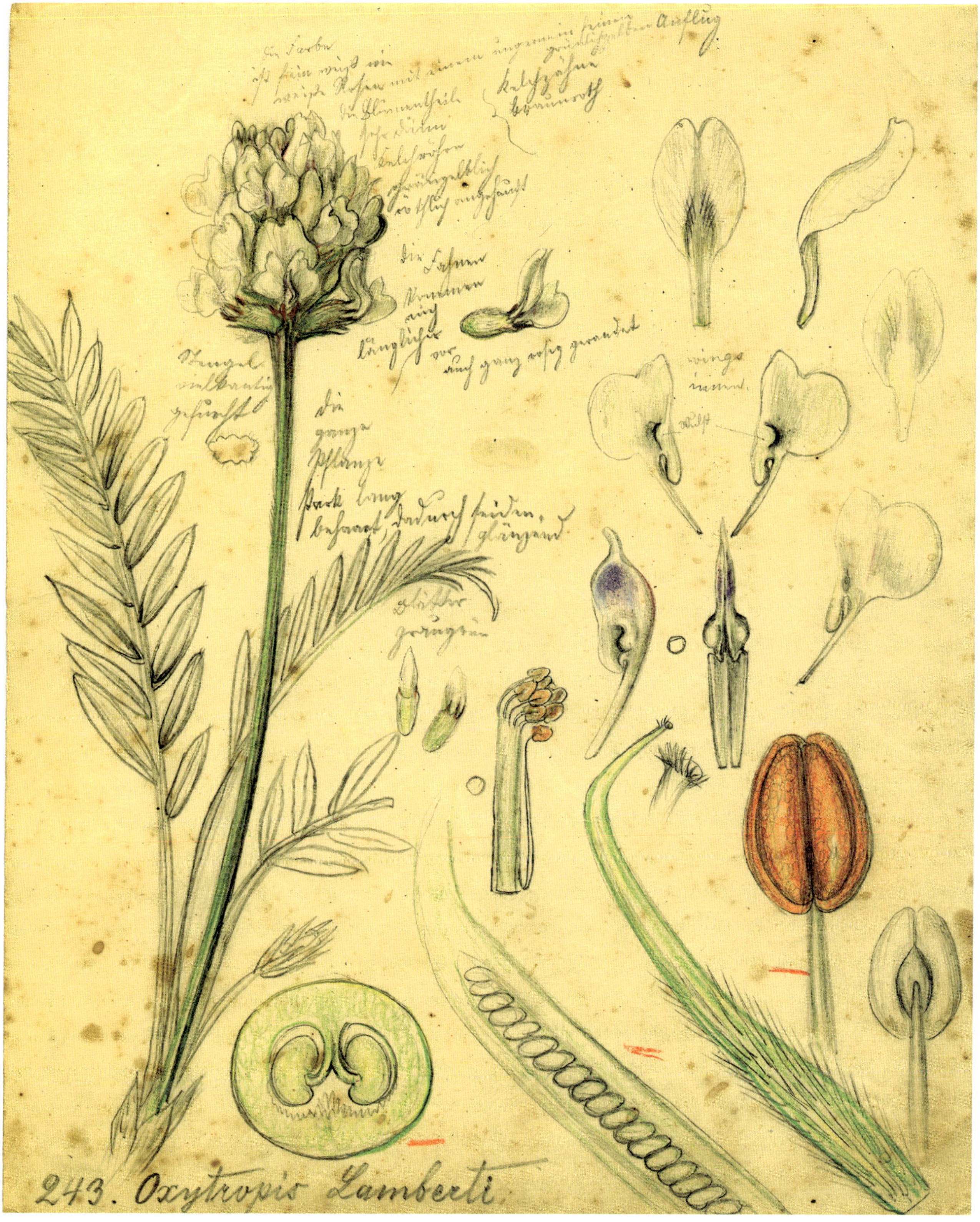
243. Oxytropis Lamberti.
wings

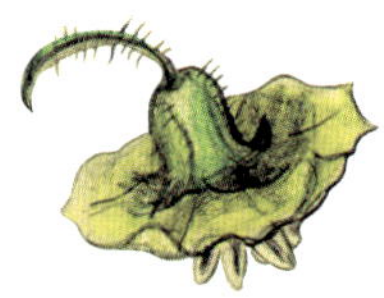

29.
Physalis lanceolata Michx.
Sword Ground-Cherry
Solanaceae
Model 434 (1894)
L. and R. Blaschka
Drawing no. 274
Colored pencil on paper
21 cm x 26.7 cm

This perennial native was collected near St. Louis, Missouri, in 1892. Rudolf Blaschka wrote that the flower's distinctive "brown pattern is somewhat transparent toward the bottom." The drooping calyx near the bottom leaves, he said, looks like a "goblet somewhat shiny with hair." The leaves, "light green with a matte finish" on top, are "lighter and shiny, also bristly with hair" underneath. The stem is "three-sided and furrowed, pretty shiny, hairy, light green," and "touched with purple" near the leaves.

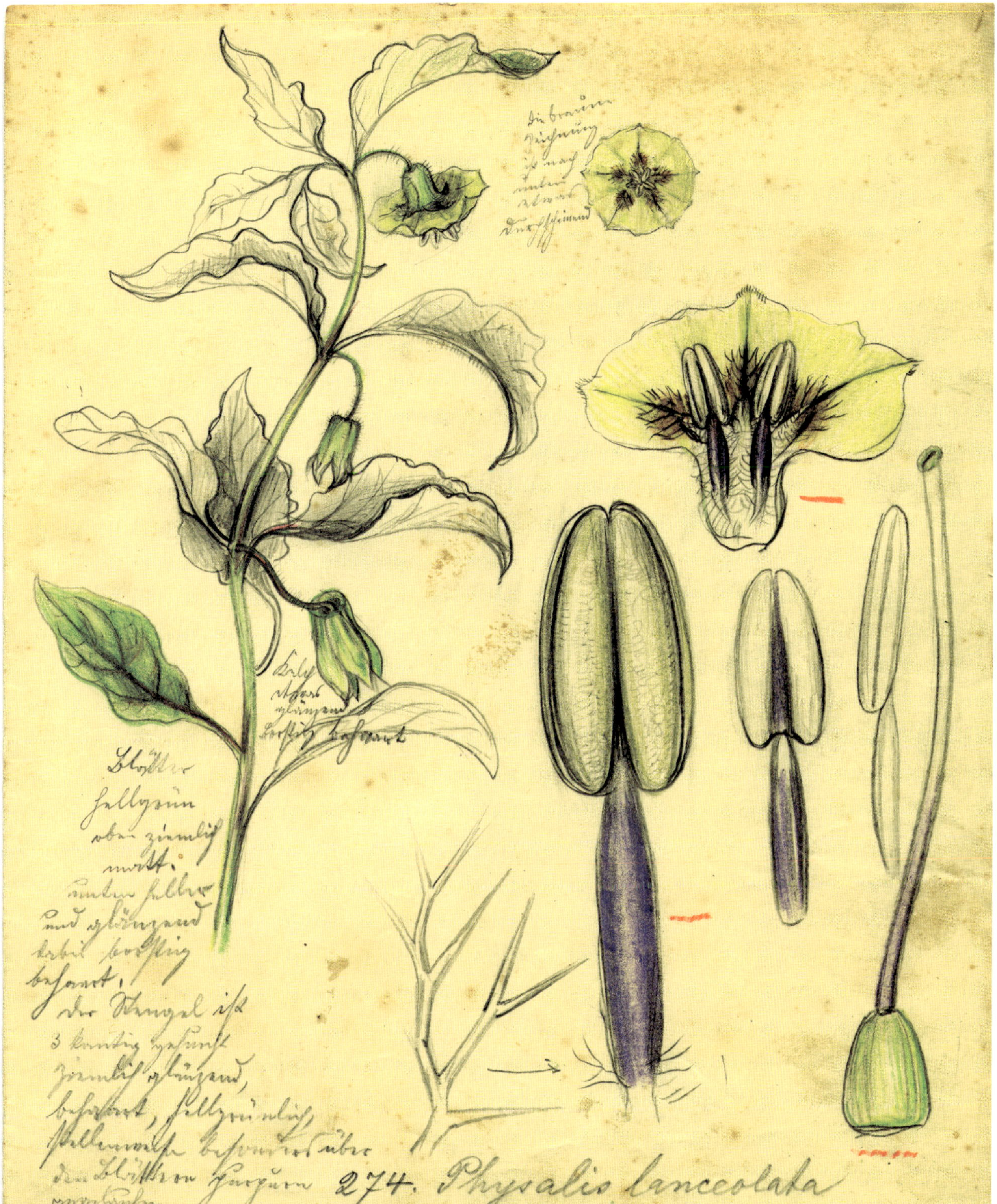
274. Physalis lanceolata

30.
Yucca filamentosa L.
Adam's Needle
Agavaceae
Model 748 (1913)
R. Blaschka
No number
Pencil on paper
16.5 cm x 21 cm

Although the Blaschkas made a yucca model in 1888, Rudolf's drawing of the microscopic stomata (specialized cells for gas exchange) on the surface of the leaf was prepared from later field observations. On April 25, 1892, he wrote, "In New Mexico and Old Mexico, I first saw the yucca desert, where, on the endless sandy plains, there is nothing other than *Yucca gloriosa, Y. filamentosa*, and other species." This is one of many comparative studies he made.

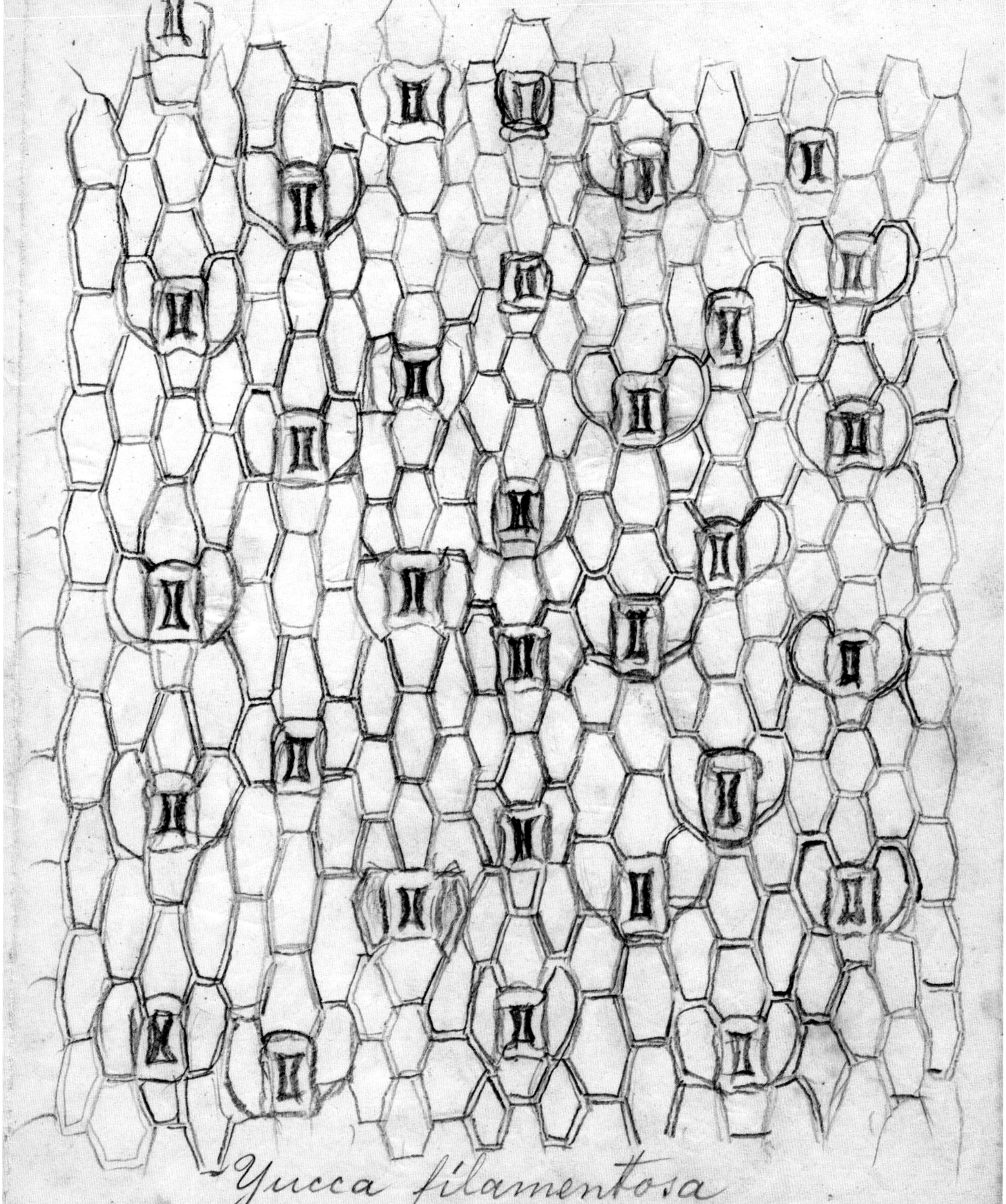
Yucca filamentosa

Further Reading

The following short list is confined to publications in English. Many of the printed works, together with information on other publications about the Blaschkas, are available at the Rakow Research Library of The Corning Museum of Glass (rakow@cmog.org).

Background

Barber, Lynn. *The Heyday of Natural History, 1820–1870.* London: Jonathan Cape, 1980.

Brunner, Bernd. *The Ocean at Home: An Illustrated History of the Aquarium.* New York: Princeton Architectural Press, 2005.

Chadarevian, Soraya de, and Nick Hopwood, eds. *Models: The Third Dimension of Science.* Stanford, California: Stanford University Press, 2004.

Hopwood, Nick. *Embryos in Wax: Models from the Ziegler Studio.* Cambridge, England: Whipple Museum of the History of Science; and Bern: Institute of the History of Medicine, 2002.

Kohlstedt, Sally Gregory. "Henry A. Ward: The Merchant Naturalist and American Museum Development." *Journal of the Society for the Bibliography of Natural History,* v. 9, 1980, pp. 647–661.

Reiling, Henri. "Václav Frič (1839–1916): Traces in Archives and Museums." *Verhandlungen zur Geschichte und Theorie der Biologie,* v. 5, 2000, pp. 341–357.

Reiling, Henri, and Tat'jána Spunarová. "Václav Frič (1839–1916) and His Influence on Collecting Natural History." *Journal of the History of Collections,* v. 17, no. 1, 2005, pp. 23–43.

Rossi-Wilcox, Susan M., Henri Reiling, and Philip Bisaga. "The Blaschkas' Lampworking Tables." *Journal of Glass Studies,* v. 45, 2003, pp. 167–176.

Sheets-Pyenson, Susan. *Cathedrals of Science. The Colonial Natural History Museums during the Late Nineteenth Century.* Kingston and Montreal: McGill-Queen's University Press, 1998.

Strong, Augustus Hopkins. *Henry A. Ward: Reminiscence and Appreciation.* Rochester, New York: The Rochester Historical Society, 1922.

Models of Invertebrate Animals

Meechan, Chris. "A Glass Menagerie: The Work of Leopold and Rudolph [*sic*] Blaschka." *The Glass Cone,* no. 39, Spring 1995, pp. 4–5.

—. "A World in Glass: Leopold and Rudolf Blaschka." In Rachel Kent, *Liquid Sea* (exhibition catalog). Sydney: Museum of Contemporary Art, 2003, pp. 20–21.

Monaghan, Nigel, and others. *Blaschka Glass Models.* Dublin: National Museum of Ireland–Natural History, 2006.

Peto, James, and Angie Hudson, eds. *Leopold & Rudolf Blaschka.* London: Design Museum, 2002.

Reiling, Henri. "The Blaschkas' Glass Animal Models: Illustrations of 19th Century Zoology." *Scientiarum Historia,* v. 26, 2000, pp. 131–143.

—. "The Blaschkas' Glass Animal Models: Origins of Design." *Journal of Glass Studies,* v. 40, 1998, pp. 105–126.

Walsche, Damien, and Nigel Monaghan. "The Collection of Blaschka Glass Models in the Natural History Museum, Dublin." *Museum Ireland,* v. 13, 2003, pp. 43–48.

Warmus, William. "More than Art: The Blaschka Marine Invertebrates." *Urban Glass,* no. 83, 2001, pp. 40–45.

Whitehouse, David. "The Blaschka Animals." *Glass,* no. 44, Summer 1991, pp. 36–41.

Models of Plants

Daston, Lorraine. "Appearances All the Way Down: The Glass Flowers as Scientific Models." In *Kunstformen des Meeres: Zoologische Glasmodelle von Leopold und Rudolf Blaschka, 1863–1890* (exhibition catalog). Tübingen, Germany: Universitätsstadt Tübingen, 2006, pp. 61–67.

Pantano, Carlo G., Susan M. Rossi-Wilcox, and David Lange. "The Glass Flowers." In *The Prehistory & History of Glassmaking Technology,* Ceramics and Civilization, v. 8, Westerville, Ohio: The American Ceramic Society, 1998, pp. 61–78.

Rossi-Wilcox, Susan M. "Art or Science? The Glass Flower Models at Harvard University." In *Kunstformen des Meeres* [see Daston], pp. 49–58.

—. "Blaschka Glass Flowers." *Glass Art,* v. 12, no. 1, November/December 1996, pp. 22–27.

—. "The Discovery of Leopold Blaschka's First Glass Models of Plants." *Journal of Glass Studies,* v. 37, 1995, pp. 145–147.

Rossi-Wilcox, Susan M., and Stuart Lindsay. "Fragile Ferns: The Blaschka Models at Harvard University." *Pteridologist,* v. 4, pt. 3, 2004, pp. 92–96.

Schultes, Richard Evans, and William A. Davis, with Hillel Burger. *The Glass Flowers at Harvard.* Cambridge, Massachusetts: Botanical Museum of Harvard University, 1992 (first published 1982).

Online Resources

http://exhibits.mannlib.cornell.edu/blaschka/more_info/index.htm ("Out of the Teeming Sea," online exhibition based on the collection of 570 models of invertebrates acquired by Cornell University).

http://www.hmnh.harvard.edu/exhibitions/glassflowers.html (The Ware Collection of Blaschka Glass Models of Plants at the Harvard Museum of Natural History).

http://www.ucd.ie/blaschka/dublin_coll.htm (the collection of 530 models of invertebrates in the National Museum of Ireland–Natural History).

Indexes

General Index

Note: Museums, universities, gardens, and other points of interest are indexed under the names of the cities in which they are located. The names of countries that follow city names vary according to the time discussed in the text. For example, "Böhmisch Aicha" is followed by "Bohemia" because that was the location of the city at the time of the Blaschkas. The current name of the city, "Český Dub," is followed by "Czech Republic." Some city names, including those that are now part of Germany, are followed by their current country names because these are the names that are used in the text.

A

B

C

D

E

F

G

H

I

J

K

L

M

N

O

P

R

S

T

U

V

W

Y

Z

Index to Drawings

SCIENTIFIC NAMES

Invertebrate Drawings

Botanical Drawings

COMMON NAMES

Invertebrate Drawings

Botanical Drawings